Agile Software Development Quality Assurance

Ioannis G. Stamelos
Aristotle University of Thessaloniki, Greece

Panagiotis Sfetsos
Alexander Technological Educational Institution of Thessaloniki, Greece

INFORMATION SCIENCE REFERENCE
Hershey · London · Melbourne · Singapore

Acquisitions Editor:	Kristin Klinger
Development Editor:	Kristin Roth
Senior Managing Editor:	Jennifer Neidig
Managing Editor:	Sara Reed
Assistant Managing Editor:	Sharon Berger
Copy Editor:	Larissa Vinci
Typesetter:	Sara Reed
Cover Design:	Lisa Tosheff
Printed at:	Yurchak Printing Inc.

Published in the United States of America by
Information Science Reference (an imprint of Idea Group Inc.)
701 E. Chocolate Avenue, Suite 200
Hershey PA 17033
Tel: 717-533-8845
Fax: 717-533-8661
E-mail: cust@idea-group.com
Web site: http://www.info-sci-ref.com

and in the United Kingdom by
Information Science Reference (an imprint of Idea Group Inc.)
3 Henrietta Street
Covent Garden
London WC2E 8LU
Tel: 44 20 7240 0856
Fax: 44 20 7379 0609
Web site: http://www.eurospanonline.com

Library of Congress Cataloging-in-Publication Data

Agile software development quality assurance / Ioannis Stamelos and Panagiotis Sfetsos, editors.

p. cm.

Summary: "This book provides the research and instruction used to develop and implement software quickly, in small iteration cycles, and in close cooperation with the customer in an adaptive way, making it possible to react to changes set by the constant changing business environment. It presents four values explaining extreme programming (XP), the most widely adopted agile methodology"--Provided by publisher.

Includes bibliographical references and index.

ISBN 978-1-59904-216-9 (hardcover) -- ISBN 978-1-59904-218-3 (ebook)

1. Computer software--Development. 2. Computer software--Quality control. I. Stamelos, Ioannis, 1959- II. Sfetsos, Panagiotis, 1953-

QA76.76.D47A394 2007

005.3--dc22

British Cataloguing in Publication Data
A Cataloguing in Publication record for this book is available from the British Library.

Table of Contents

Section III
Quality within Agile Process Management

Section IV
Agile Methods and Quality: Field Experience

Detailed Table of Contents

This chapter provides a review of the state-of-the-art of agile methodologies. However, it focuses primarily on the issue of quality and quality assurance, reviewing the benefits that agile methods have brought to software development. An analysis framework is used for systematically analyzing and comparing agile methodologies and is applied to three of them.

Following the presentation of the previous chapter, the agile information systems development process is discussed here and its quality characteristics are analyzed in detail. One important issue is raised: how suitable and applicable are agile methods when applied on different organisational and national situations? The text provides arguments on the basis of the authors' experiences from various European countries differing in their academic and work values, and information systems development industrial practices.

In this chapter, arguments are provided in favour of the quantification of agile processes to reinforce quality assurance procedures. Measuring requirements, design artefacts, and delivered results provide the basis for sound quality estimation. The text discusses in detail the benefits of quantification and proposes the quantification approach Planguage. Interesting results from Planguage application in the context of a Norwegian organization are given.

Section II
Quality within Agile Development

In this chapter, the authors describe a number of approaches for managing user requirements (namely software requirements specification, use cases, interaction design scenarios). Requirements are subject to constant changes in modern software development and the text shows how agile methods promote the involvement of customers/users in the process of requirement modification. The tool for assuring requirements quality are user stories and is thoroughly discussed and illustrated in this chapter.

This chapter discusses refactoring, an agile procedure during which, among other activities, quality defect removal takes place. Because of time constraints, quality defects can not be removed in just one refactoring phase. Documentation of detected quality defects is therefore necessary and the text proposes a process for the recurring and sustainable discovery, handling, and treatment of quality defects in software systems. The process is based on an annotation language, capable to register information about quality defects found in source code.

This chapter proposes a process-based approach for assuring quality while developing in agile mode. The authors propose a new concentric loop-based technique, which effectively utilizes resources during iterative development. It is based on three types of testing, namely crash testing, smoke testing, and comprehensive testing. The overall approach is illustrated on the development of graphical user interfaces. The GUI model used to implement the concentric-loop technique is given in detail.

Section III
Quality within Agile Process Management

Because of the frequent changes, multiple iterations, and software versions that occur in agile development, software configuration management is a crucial activity. This chapter discusses the additional requirements for software configuration management with respect to the traditional development. Typical agile activities for configuration management are described along with general guidelines. It is argued that an agile project can assure better quality according to the agile method and configuration management it applies and the project particular context.

Chapter VIII
Improving Quality by Exploiting Human Dynamics in Agile Methods / *Panagiotis Sfetsos and Ioannis Stamelos* 154

This chapter explores the management of the human resources that are involved in agile development. Because evidently human factors are critical for the success of agile methods, there is an urgent need for managing agile people effectively both at the corporate level and the project level. First part of the chapter proposes and discusses a model for personnel management based on the well-known People-CMM assessment and improvement model. In addition, the chapter proposes a model for allocating and rotating developers in pairs while pair programming. The model is based on the fact that different types of personalities and temperaments allow pairs that produce better quality results.

Chapter IX
Teaching Agile Software Development Quality Assurance / *Orit Hazzan and Yael Dubinsky*.......... 171

This chapter differs from the rest of the book in the sense that it deals with the education of software engineers and managers to form a culture for agile quality assurance. The text proposes a teaching framework focusing on the way quality issues are perceived in agile software development environments. It consists of nine principles, which can be adjusted according to different specific teaching environments. The teaching framework is based on the differences between agile and traditional software development. Overall, this chapter provides a particularly useful tool for instructors of Agile Methods.

Section IV
Agile Methods and Quality: Field Experience

Chapter X
Agile Software Development Quality Assurance: Agile Project Management, Quality Metrics, and Methodologies / *James F. Kile and Maheshwar K. Inampudi*.. 186

This chapter examines one of the hottest issues in modern software development, namely the adoption of agile methods by highly disciplined and highly structured software development environments. It appears that up to now, agile methods have been applied mostly to non-critical projects. The text describes how one IBM software development team has applied simultaneously several individual agile development techniques. The authors report encouraging results, stating that they obtained increased quality in shorter than normal time. Throughout the chapter, it is shown that the adoption of individual agile techniques can be achieved with no additional risks.

This chapter describes the practice of test-driven development and the benefits it brings to quality assurance in an agile organization. The practice is illustrated through details of two real development projects in an industrial setting. The author gives an industry practitioner's perspective and discusses various practical considerations about the adoption of the practice. Overall, it is claimed that test-driven development is well accepted by practitioners and is a successful quality assurance technique.

In this chapter, the experience of another large company, namely Siemens, with agile methodologies is reported. The authors report that Siemens has applied agile processes in several projects with varying characteristics. They also report that significant quality achievements have been observed. The text discusses briefly project quality goals and practices and summarizes the lessons learned from successes and failures while working for quality assurance in their projects. This chapter is important because it shows how a large company pursues quality assurance results when applying agile methods.

Foreword

After spending the summer north of the Arctic Circle, basking in the midnight sun and the warmest weather for over 100 years in Finland, I was especially happy to find this book sitting on my desk waiting to be read. Although there is no shortage of books on agile methodologies and practices, something had been missing. The concept of quality is indeed a very important element in any software system and development method, yet it has received little explicit attention in the agile literature. For this reason, I am delighted to see this book contribute to this gap.

We have long known that skilled people are the most crucial resource in software development. Back in the 1990 summer issue of *American Programmer* (Ed Yourdon's Software Journal, Vol. 3, No. 7-8)—which was devoted exclusively to "Peopleware"—the editor commented that "Everyone knows the best way to improve software productivity and quality is to focus on people." However, it took more than 10 years for the agile manifesto and agile methods (Extreme Programming, Scrum, Crystal, and many others) to truly place the emphasis on people and their interaction. Since then, we have witnessed a movement that has advanced more rapidly than any other innovation in the field of software engineering.

Software quality in agile development is not a straightforward topic. Therefore, it is essential that a book of this kind does not aim at offering simple answers to complex problems. An edited book allows the contributors to approach the topic from their particular angles in an in-depth manner. In this book there are chapters not normally found in the agile literature dealing with, for example, metrics and documenting defects. Some of the chapters take a controversial approach and offer new insights into adapting agile methods in different development situations. The reader will quickly realise that these types of arguments, studies, and suggestions are much needed in this field.

The reader can further enjoy the opportunity to select and read the contents pertaining to their background and interests. I am happy to see that the editors have succeeded in collecting chapters that not only build upon one another but, more importantly, form a coherent whole addressing the relevant issues from people management to coding with experiences drawn from the industry. And all of this is addressed from the perspective of software quality!

As an academic, I value the fact that this book includes a number of rigorously performed scientific studies. This is particularly welcome as it enables us to answer the question why agile methods work. To date, we have seen quite interesting anecdotal evidence that agile methods do improve quality and even make the programmers' work a happier one. However, this book contributes also to the scientific discussion by providing thoughts and theories that explain the results.

Sometimes we tend to forget that one of the better ways to influence the future of software development is to offer specific material for teachers who educate young developers in universities and other educational institutes. While I believe all the chapters are of merit in this book, I am impressed to find a chapter written for the use of educators as well.

Whether you read this book from start to finish, or piecemeal your approach iteratively, I am sure you will find this book as valuable as I did.

Pekka Abrahamsson
Research Professor
VTT Technical Research Centre of Finland

Pekka Abrahamsson is a research professor at VTT Technical Research Centre of Finland. Currently, he is on leave from the University of Tampere where he is a full professor in the field of information systems and software engineering. His current responsibilities include managing an AGILE-ITEA project (http://www.agile-itea.org), which involves 22 organizations from nine European countries. The project aims at developing agile innovations in the domain of complex embedded systems. His research interests are centred on mobile application development, business agility, agile software production, and embedded systems. He leads the team who has designed an agile approach for mobile application development—the Mobile-D. He has coached several agile software development projects in industry and authored 50+ scientific publications focusing on software process and quality improvement, agile software development and mobile software. His professional experience involves 5 years in industry as a software engineer and a quality manager

Preface

Agile methods drastically alter the software development processes. Agile software processes, such as extreme programming (XP), Scrum, etc., rely on best practices that are considered to improve software development quality. It can be said that best practices aim to induce software quality assurance (SQA) into the project at hand. Proponents of agile methods claim that because of the very nature of such methods, quality in agile software projects should be a natural outcome of the applied method. As a consequence, agile software development quality assurance (ASDQA) is hoped/expected/supposed to be more or less embedded in the agile software processes, while SQA practices are integrated across the entire life-cycle development, from requirements through the final release. Thus, agile methods introduce a different perspective on QA in software development.

Agile practices are expected to handle unstable and volatile requirements throughout the development lifecycle, to deliver software with fewer defects and errors, in shorter timeframes, and under predefined budget constraints. The iterative and incremental way of development allows both customer requirements revision mechanisms and customer active participation in the decision-making process. Customer participation provides the needed feedback mechanism, ensuring customer perceived satisfaction for the final product. It is also known that agile methods make the key business users a very strong partner in assuring quality. Rather than completely leaving quality to the professionals, agile projects make these key users responsible for ensuring that the application is fit for purpose. Agile development embraces test driven development and test first design, both coming from the arena of good practices, introducing them into mainstream development, and minimizing errors and defects of the final product. Some other practises, such as simple planning and designing, pair programming, short iteration cycles, small releases, continuous integrations, common code ownership, and metaphor potentially reinforce quality assurance.

It is interesting to note that the previously mentioned practices cover and support, to a significant extent, total quality management (TQM) (see Crosby, 1979; Deming, 1986; Feigenbaum, 1961, 1991; Ishikawa, 1985; Juran & Gryna, 1970, all referenced in Chapter II). We remind the reader that a TQM system comprises four key common elements: (1) customer focus, (2) process improvement, (3) human side of quality, and (4) measurement and analysis. Agile methods deal in one way or another with all four elements. Many reports support and evangelize the advantages of agile methods with respect to quality assurance, even if the term "quality assurance" is avoided as coming from traditional, bureaucratic development.

Is it so? For example, is it the case that agile methods assure quality by default, and software managers/developers need not be concerned with quality issues, such as quality planning, quality audits, or quality reports? Proponents of agile methods must provide convincing answers to questions such as "What is the quality of the software produced?" or "Which hard/soft evidence supports the superiority of agile quality?" There has been little published work that focuses on such agile software development quality issues. In particular, there is a literature gap in providing a critical view of agile quality, pinpointing areas where agile methods are strong, but also areas that need improvement.

OVERALL OBJECTIVE OF THE BOOK

This book pursues an ambitious goal: it attempts to provide answers to the questions and issues previously raised. It provides original academic work and experience reports from industry related to agile software development quality assurance. Its mission is to describe fundamentals of ASDQA theory and provide concrete results from agile software development organizations. To understand how quality is or should be handled, the whole development process must be analyzed, measured, and validated from the quality point of view, as it is claimed to be the rule when traditional methods are employed. It is precisely from the quality point of view that the book looks at agile methods. The area is wide and entails many facets that the book attempts to clarify, including:

- Differences and similarities between the traditional quality assurance procedures and ASDQA.
- Identification and evaluation of quality metrics in agile software development.
- Reports on the state of the art regarding quality achievements in agile methods.
- Investigation on how practices and tools affect the quality in agile software development.
- Human issues in ASDQA.
- Education in ASDQA concepts and techniques.

Book chapters provide theoretical discussion on ASDQA issues and/or results and lessons from practical ASDQA application. Eventually, the book is expected to provide successful quality management tips that can help participants in the agile software development process avoid risks and project failures that are frequently encountered in traditional software projects. Because such task is extremely difficult, given the variety of agile methods, the relatively limited time they have been exercised and the scattered, often vague, information regarding agile quality from the field, this book could only be edited, and not be written by a small authors' group.

The book takes the form of a collection of edited chapters. Authors of the chapters cover all kinds of activities related to agile methods: they are academicians, practitioners, consultants, all involved heavily in practicing, researching, and teaching of agile methods. Authors come from almost all over the world (North America, Europe, Asia, Africa) and are employed by all kinds of organizations involved in agile development (universities, research institutes, small or large agile development/consulting companies).

ORGANIZATION OF THE BOOK

This book is made up of 12 chapters, organized in four sections. Section titles are the following:

Section I: Introduction: Agile Methods and Quality
Section II: Quality within Agile Development

Section III: Quality within Agile Process Management
Section IV: Agile Methods and Quality: Field Experience

Section I: Introduction: Agile Methods and Quality provides the framework for the rest of the book. It is particularly useful for readers not familiar with all aspects of agile methods. It reviews agile methods and compares them with traditional approaches. Section I starts posing questions about the quality achieved and potential problems with agile methods today. It also starts to propose solutions for certain identified issues.

Section II: Quality within Agile Development examines how quality is pursued throughout software development. It gives a flavour of how developers achieve quality in an agile fashion. Chapters in this section review quality assurance when specifying requirements, when handling defects, and when user interfaces are designed and implemented.

Section III: Quality within Agile Process Management examines how quality is pursued throughout the handling of agile software processes. This section deals with activities that run parallel to development or prepare the development teams for effective work. It gives a flavour of how managers achieve quality in an agile fashion. Two chapters in this Section review quality assurance when managing agile software configurations and when agile people are managed. Finally, a critical theme for the future is addressed, namely the education of next generations of agile developers and managers in ASDQA issues.

Section IV: Agile Methods and Quality: Field Experience provides feedback from agile method application. Although all chapters up to now try to capture experiences from agile projects and to incorporate them in theoretical frameworks, chapters of this section come right from agile companies. Interestingly, two of the Chapters come from quite large companies, signalling the expansion of agile methods into the realm of traditional software development. Chapters provide invaluable information about agile project management, quality measurement, test driven development and, finally, lessons learned from ASDQA real world application.

A brief description of each chapter follows. Chapters are organized according to the sections they belong.

Section I: Introduction: Agile Methods and Quality

Chapter I: Agile Software Methods: State-of-the-Art

In Chapter I, Ernest Mnkandla and Barry Dwolatzky (South Africa) analyze and define agile methodologies of software development. They do so by taking a software quality assurance perspective. The chapter starts by defining agile methodologies from three perspectives: a theoretical definition, a functional definition, and a contextualized definition. Next, a brief review of some of the traditional understandings of quality assurance is given, and the author proceeds with certain innovations that agility has added to the world of quality. Doing so, the text provides an understanding of the state-of-the-art in agile methodologies and quality, along with expectations for the future in this field. An analysis framework is used for objectively analyzing and comparing agile methodologies. The framework is illustrated by applying it to three specific agile methodologies.

Chapter II: Agile Quality or Depth of Reasoning? Applicability vs. Suitability with Respect to Stakeholders' Needs

In Chapter II, Eleni Berki (Finland), Kerstin Siakas (Greece), and Elli Georgiadou (UK) provide an in-depth discussion and analysis of the quality characteristics of the agile information systems development process. They question ASDQA by exposing concerns regarding the applicability and suitability of agile methods in different organisational and national cultures. They argue based on recent literature reviews and published reports on the state-of-the-art in agile Methodologies. A unique feature of this chapter is that its authors draw their experience from different European countries (Denmark, England, Finland, Greece) with diverse academic and work values, and information systems development (ISD) industrial practices based on different principles. They relate and compare traditional, agile, managed, and measured ISD processes, they explore human dynamics that affect success and consensus acceptance of a software system and propose a critical framework for reflecting on the suitability and applicability of agile methods in the development and management of quality software systems. To achieve this, the authors examine the different European perceptions of quality in the agile paradigm and compare and contrast them to the quality perceptions in the established ISD methodological paradigms.

Chapter III: What's Wrong with Agile Methods? Some Principles and Values to Encourage Quantification

In Chapter III, Tom Gilb (Norway) proposes the quantification of agile processes to reinforce ASDQA. He claims that agile methods could benefit from using a more quantified approach across the entire implementation process (that is, throughout development, production, and delivery). He discusses such things as quantification of the requirements, design estimation, and measurement of the delivered results. He outlines the main benefits of adopting such an approach, identifying communication of the requirements, and feedback and progress tracking as the areas that are most probable to benefit. The chapter presents the benefits of quantification, proposes a specific quantification approach (Planguage), and finally describes a successful case study of quantifying quality in a Norwegian organization.

Section II: Quality within Agile Development

Chapter IV: Requirements Specification user Stories

In this chapter, Vagelis Monochristou and Maro Vlachopoulou (Greece) review quality assurance in the requirements specification development phase. Such phase is known to give a lot of problems and injects hard to detect and correct defects in the documentation and the software itself. The authors discuss several approaches, which suggest ways of managing user's requirements (software requirements specification, use cases, interaction design scenarios, etc.). They emphasize the fact that many real users requirements appear in development phases following the initial ones. One way to cope with this situation is to involve customers/users in these development phases as well. When provided with insight about the various sub-systems as they are developed, customers/users can re-think and update their requirements. However, to accommodate such customer/user role within the development cycle, software organizations must take a non-traditional approach. Agile methods are this alternative approach because of the iterative and incremental way of development they propose. Allowing for iteration and gradual system building, user requirements revision mechanisms, and active user participation is encouraged and supported throughout the development of the system. User stories are the agile answer to the problem and they are thoroughly discussed and illustrated in this chapter.

Chapter V: Handling of Software Quality Defects in Agile Software Development

Although the previous chapter told us how to capture and avoid problems in user requirements, defects can still be injected in the software code. In agile software development and maintenance, the phase that allows for continuous improvement of a software system by removing quality defects is refactoring. However, because of schedule constraints, not all quality defects can be removed in just one refactoring phase. Documentation of quality defects that are found during automated or manual discovery activities (e.g., pair programming) is necessary to avoid waste of time by rediscovering them in later phases. However, lack of documentation and handling of existing quality defects and refactoring activities is a typical problem in agile software maintenance. In order to understand the reason for modifying the code, one must consult either proprietary documentations or software versioning systems. Jörg Rech (Germany), the author of this chapter, describes a process for the "recurring and sustainable discovery, handling, and treatment of quality defects in software systems." His proposed tool for assuring quality in this context is an annotation language, capable to register information about quality defects found in source code, representing the defect and treatment activities of a software sub-system. One additional benefit from using such annotation language is that it can also be useful during testing and inspection activities.

Chapter VI: Agile Quality Assurance Techniques for GUI-Based Applications

In this chapter, Atif Memon and Qing Xie (USA) adopt a strong, process-based approach for assuring quality while developing in agile mode. They discuss the need for new agile model-based testing mechanisms, neatly integrated with agile software development/evolution and propose a new concentric loop-based technique, which effectively utilizes resources during iterative development. They call the inner loop "crash testing," applied on each code check-in of the software. The second loop is called smoke testing and operates on each day's build. The outermost loop is called the "comprehensive testing" loop, executed after a major version of the software is available. The authors illustrate their approach on a critical part of today software systems, namely graphical user interface (GUI). They choose GUI front-ends because GUI development is quite suitable for agile development and because rapid testing of GUI-based systems is particularly challenging. They describe in detail the GUI model used to implement the concentric-loop technique.

Section III: Quality within Agile Process Management

Chapter VII: Software Configuration Management in Agile Development

Chapters in this section focus on project activities that are parallel to development and software configuration management (SCM) is an essential part of any software process. Because of frequent changes, multiple iterations and software versions, SCM is of particular importance for any agile project. In this chapter, Lars Bendix and Torbjörn Ekman (Sweden) discuss the peculiarities of agile SCM and argue that SCM needs to be done differently and in a more extended fashion than during traditional development. They also define the ways in which quality is assured through the application of SCM. To do so, they first provide a brief introduction to the focal SCM principles and list a number of typical agile activities related to SCM. Next, they explain the reader that it is possible to define certain general SCM guidelines for how to support and strengthen these typical agile activities. They describe the characteristics of an agile method that are necessary in order to take full advantage from SCM and, as a consequence, to better assure quality. Following the proposed guidelines, any agile project can obtain the best result from SCM according to the agile method it applies and the project particular context.

Chapter VIII: Improving Quality by Exploiting Human Dynamics in Agile Methods

This chapter deals with a completely different process issue than previous chapter, namely the management of the human resources that are involved in agile development. Panagiotis Sfetsos and Ioannis Stamelos (Greece) argue that human factors are still critical for the success of software engineering in general. In particular, agile methods are even more sensitive to human factors because they are heavily based on the contribution and effort of the individuals working in the agile project. Documentation is limited with respect to traditional development and effective inter-personal communication is necessary for successful project completion. The authors describe how a large agile organization can cope with human resource management both at the corporate level and the project level. First part of the chapter proposes and discusses a model for personnel management based on the well-known People-CMM assessment and improvement model. The agile organization can pursue higher model levels by assessing its current situation and by introducing advanced human resource management practices. In doing so, the organization must take profit from the distinguished way in which humans are involved in agile methods and activities. Second part proposes a model that exploits developer personalities and temperaments to effectively allocate and rotate developers in pairs for pair programming. The rationale is that by mixing different types of personalities and temperaments, pairs become more productive and quality is more easily assured.

Chapter IX: Teaching Agile Software Development Quality Assurance

This chapter ends the section on agile process issues dealing with the preparation of software engineers and managers to address agile quality assurance. Authors Orit Hazzan and Yael Dubinsky (Israel) provide a teaching framework that focuses on the way quality issues are perceived in agile software development environments. The teaching framework consists of nine principles, which can be adjusted according to different specific teaching environments and therefore implemented in various ways. The chapter outlines these principles and addresses their contribution to learners' understanding of agile quality. The authors enrich the discussion of their teaching framework by identifying the differences between agile and traditional software development in general, and with respect to software quality in particular. The material of the chapter can be used by software engineering instructors who wish to base students learning on students' experience of the different aspects involved in software development environments.

Section IV: Agile Methods and Quality: Field Experience

Chapter X: Agile Software Development Quality Assurance: Agile Project Management, Quality Metrics, and Methodologies

In the first chapter of the section with results end experiences from agile companies, James F. Kile and Maheshwar R. Inampudi (IBM, USA) deal with a really hot issue, crucial for the further expansion of agile methods. They ask whether "the adaptive methods incorporated within many of the most popular agile software development methodologies can be successfully implemented within a highly disciplined and highly structured software development environment and still provide the benefits accorded to fully agile projects." They observe that agile methods have been applied mostly to non-critical projects, by small project teams, with vague requirements, a high degree of anticipated change, and no significant availability or performance requirements. It is therefore questionable whether agile methods can be applied in situations with strong quality requirements. The authors report an extremely interesting experience: they describe how one team adopted not one single agile method, but several individual agile development techniques. They manage to achieve software development quality improvements, while in parallel reducing overall cycle time. The authors propose that all is needed is a common-sense approach to software development. Overall, they demonstrate that the incorporation of individual agile techniques may be done in such a way that no additional risk is incurred for projects having high availability, performance, and quality requirements.

Chapter XI: Test-Driven Development: An Agile Practice to Ensure Quality is Built from the Beginning

This chapter is written by Scott Mark (Medtronic, USA) and describes the practice of test-driven development (TDD) and its impact on the overall culture of quality and quality assurance in an organization. The discussion on this popular practice is based on the author's personal experience introducing TDD into two existing development projects in an industrial setting. He discusses basic concepts of TDD from an industry practitioner's perspective and he proceeds with an elaboration of the benefits and challenges of adopting TDD within a development organization. He reports to the reader that TDD was well-received by team members, and he is optimistic, in the sense that other teams will behave in the same manner, provided that they are prepared to evaluate their own experiences and address the challenges imposed by TDD.

Chapter XII: Quality Improvements from using Agile Development Methods: Lessons Learned

This chapter, ending the session with experiences from industry (and the book), comes from another large company, namely Siemens (USA). Beatrice Miao Hwong, Gilberto Matos, Monica McKenna, Christopher Nelson, Gergana Nikolova, Arnold Rudorfer, Xiping Song, Grace Yuan Tai, Rajanikanth Tanikella, and Bradley Wehrwein report that "in the past few years, Siemens has gained considerable experience using agile processes with several projects of varying size, duration, and complexity." The authors build on this invaluable experience for the agile world and report that they have observed "an emerging pattern of quality assurance goals and practices across these experiences." They describe the projects in which they have used agile processes. They also provide information on the processes themselves. They discuss briefly project quality goals and practices and present (as the chapter title promises) the lessons learned from the successes and failures in practicing quality assurance in agile projects. The material they provide is informative about the methods they employed for achieving the established quality goals, leading to a first-hand understanding of the current state of ASDQA.

Acknowledgments

This book has become a reality only because of the hard work of the chapter authors. We sincerely wish to thank them for the time they devoted to write their chapter, peer review two chapters of co-authors, and revise their chapters according to the comments they received. We also wish to thank Dr. Sulayman Sowe (Aristotle University) for proofreading a significant part of the book text. Finally, we would like to thank Kristin Roth (IGI) for her guidance and useful suggestions throughout the preparation of the material, and Idea Group Publishing for the opportunity they gave us to edit this interesting book.

Ioannis Stamelos
Panagiotis Sfetsos
Editors

About the Editors

Ioannis Stamelos is an assistant professor of computer science at the Aristotle University of Thessaloniki, Department of Informatics. He received a degree in electrical engineering from the Polytechnic School of Thessaloniki (1983) and a PhD degree in computer science from the Aristotle University of Thessaloniki (1988). He teaches language theory, object-oriented programming, software engineering, software project management, and enterprise information systems at the graduate and postgraduate level. His research interests include empirical software evaluation and management, software education, and open source software engineering. Stamelos is the author of 60 scientific papers and a member of the IEEE Computer Society.

Panagiotis Sfetsos is a lecturer of computer science at Alexander Technological Education Institute of Thessaloniki, Greece, Department of Informatics, since 1990. His research interests include experimentation in SE, agile methods, extreme programming, software measurement, software testing, and quality. Sfetsos received his BSc in computer science and statistics from the University of Uppsala, Sweden (1981), and then worked for several years in software development at industry and education. He published a

Section I
Introduction:
Agile Methods and Quality

Chapter I
Agile Software Methods: State-of-the-Art

Ernest Mnkandla
Monash University, South Africa

Barry Dwolatzky
University of Witwatersrand, South Africa

ABSTRACT

This chapter is aimed at comprehensively analyzing and defining agile methodologies of software development from a software quality assurance perspective. A unique way of analyzing agile methodologies to reveal the similarities that the authors of the methods never tell you is introduced. The chapter starts by defining agile methodologies from three perspectives: a theoretical definition, a functional definition, and a contextualized definition. Then an agile quality assurance perspective is presented starting from a brief review of some of the traditional understandings of quality assurance to the innovations that agility has added to the world of quality. The presented analysis approach opens a window into an understanding of the state-of-the-art in agile methodologies and quality, and what the future could have in store for software developers. An understanding of the analysis framework for objectively analyzing and comparing agile methodologies is illustrated by applying it to three specific agile methodologies.

INTRODUCTION

Agile software development methodologies have taken the concepts of software quality assurance further than simply meeting customer requirements, validation, and verification. Agility innovatively opens new horizons in the area of software quality assurance. A look at the agile manifesto (Agile Alliance, 2001) reveals that agile software development is not just about meeting customer requirements (because even process-driven methodologies do that), but it is about meeting the changing requirements right up to the level of product deployment. This chapter introduces a technique for analyzing agile methodologies in a way that reveals the fundamental similarities among the different agile processes.

As for now, there is a reasonable amount of literature that seeks to describe this relatively new set of methodologies that have certainly changed the way software development is done. Most of the existing work is from the authors of the methodologies and a few other practitioners. What lacks is therefore a more balanced evaluation comparing what the original intents of the authors of agile methodologies were, to the actual things that have been done through agile methodologies over the last few years of their existence as a group, and the possible future applications.

While most of those who have applied agile methods in their software development projects have gained margins that are hard to ignore in the areas of product relevance (a result of embracing requirements instability) and quick delivery (a result of iterative incremental development), some have not joined this new fun way to develop software due to a lack of understanding the fundamental concepts underlying agile methodologies. Hence, this chapter intends to give the necessary understanding by comprehensively defining agile methodologies and revealing how agile methodologies have taken software quality assurance further than traditional approaches. The second concern resulted from more than three years of research into agile methodology practices where the author discovered that the individual agile methods such as extreme programming, scrum, and lean development etc. are not that different from each other. The apparent difference is because people from different computing backgrounds authored them and happen to view the real world differently. Hence, the differences are not as much as the authors would like us to believe. The evaluation technique introduced here will reveal the similarities in a novel way and address the adoption concerns of agile methodologies. This also reveals what quality in an agile context means.

CHAPTER OBJECTIVES

The objective of this chapter is to introduce you to the fundamentals of analyzing agile methodologies to reveal the bare bones of agile development. After reading this chapter, you will:

- Understand three approaches to the definition of agile methodologies (i.e., a theoretical definition, a functional definition, and a contextualized definition).
- Understand the state-of-the-art in agile methodologies.
- Understand the presented framework for objectively analyzing and comparing agile methodologies.
- Understand the meaning of software quality assurance in an agile context.

BACKGROUND

This section will start by defining agile methodologies based on what people say about agile methodologies, what people do with agile methodologies, and what agile methodologies have done to the broad area of software development.

DEFINING AGILE METHODOLOGIES

The agile software development methodologies group was given the name "agile" when a group of software development practitioners met and formed the Agile Alliance (an association of software development practitioners that was formed to formalize agile methodologies) in February 2001. The agile movement could mark the emergence of a new engineering discipline (Mnkandla & Dwolatzky, 2004a) that has shifted the values of the software development process from the mechanistic (i.e., driven by process and rules of science) to the organic (i.e., driven by softer issues of people and their interactions).

This implies challenges of engineering complex software systems in work environments that are highly dynamic and unpredictable.

THEORETICAL DEFINITION

After the first eWorkshop on agile methodologies in June 2002, Lindvall et al. (2002) summarized the working definition of agile methodologies as a group of software development processes that are iterative, incremental, self-organizing, and emergent. The meaning of each term in the greater context of agility is shown next.

1. **Iterative:** The word iterative is derived from iteration which carries with it connotations of repetition. In the case of agile methodologies, it is not just repetition but also an attempt to solve a software problem by finding successive approximations to the solution starting from an initial minimal set of requirements. This means that the architect or analyst designs a full system at the very beginning and then changes the functionality of each subsystem with each new release as the requirements are updated for each attempt. This approach is in contrast to more traditional methods, which attempt to solve the problem in one shot. Iterative approaches are more relevant to today's software development problems that are characterized by high complexity and fast changing requirements. Linked with the concept of iterations is the notion of incremental development, which is defined in the next paragraph.
2. **Incremental:** Each subsystem is developed in such a way that it allows more requirements to be gathered and used to develop other subsystems based on previous ones. The approach is to partition the specified system into small subsystems by functionality and add a new functionality with each new release. Each release is a fully tested usable subsystem with limited functionality based on the implemented specifications. As the development progresses, the usable functionalities increase until a full system is realized.
3. **Self-organizing:** This term introduces a relatively foreign notion to the management of scientific processes. The usual approach is to organize teams according to skills and corresponding tasks and let them report to management in a hierarchical structure. In the agile development setup, the "self-organizing" concept gives the team autonomy to organize itself to best complete the work items. This means that the implementation of issues such as interactions within the team, team dynamics, working hours, progress meetings, progress reports etc. are left to the team to decide how best they can be done. Such an approach is rather eccentric to the way project managers are trained and it requires that the project managers change their management paradigm all together. This technique requires that the team members respect each other and behave professionally when it comes to what has been committed on paper. In other words management and the customer should not get excuses for failure to meet the commitment and there should be no unjustified requests for extensions. The role of the project manager in such a setup is to facilitate the smooth operation of the team by liaising with top management and removing obstacles where possible. The self-organizing approach therefore implies that there must be a good communication policy between project management and the development team.
4. **Emergent:** The word implies three things. Firstly, based on the incremental nature of the development approach the system is allowed to emerge from a series of increments. Secondly, based on the self-organiz-

Figure 1. Definition of agility © copyright Ernest Mnkandla PhD thesis University of the Witwatersrand

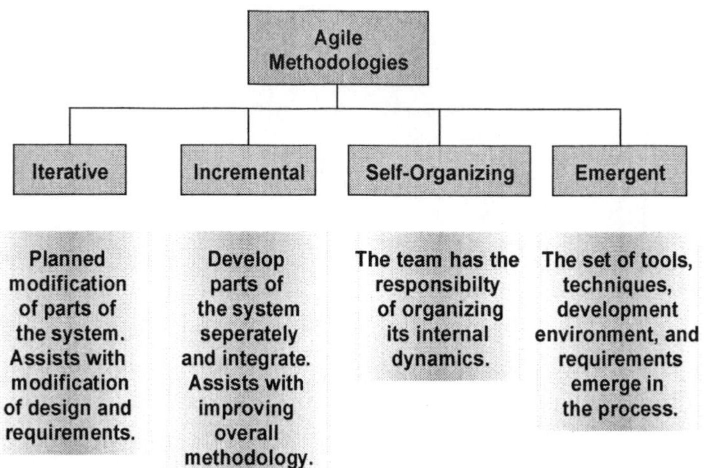

ing nature a method of working emerges as the team works. Thirdly, as the system emerges and the method of working emerges a framework of development technologies will also emerge. The emergent nature of agile methodologies means that agile software development is in fact a learning experience for each project and will remain a learning experience because each project is treated differently by applying the iterative, incremental, self-organizing, and emergent techniques. Figure 1 sums up the theoretical definition of agile methodologies.

The value of agility is in allowing the concepts defined above to mutate within the parameters set by the agile values and principles (For details on agile values and principles see the agile manifesto at http://www.agilealliance.org.. There is always a temptation to fix a framework of software development if success is repeatedly achieved, but that would kill the innovation that comes with agile development.

FUNCTIONAL DEFINITION

Agile methodologies will now be defined according to the way some agile practitioners have understood them as they used them in real world practice.

The term "agile" carries with it connotations of flexibility, nimbleness, readiness for motion, activity, dexterity in motion, and adjustability (Abrahamsson, Salo, Ronkainen, & Warsta, 2002). Each of these words will be explained further in the context of agility in order to give a more precise understanding of the kinds of things that are done in agile development.

- **Flexibility:** This word implies that the rules and processes in agile development can be easily bended to suit given situations without necessarily breaking them. In other words, the agile way of developing software allows for adaptability and variability.
- **Nimbleness:** This means that in agile software development there must be quick delivery of the product. This is usually done through the release of usable subsystems

within a period ranging from one week to four weeks. This gives good spin-offs as the customer will start using the system before it is completed.

- **Readiness for motion:** In agile development, the general intention is to reduce all activities and material that may either slow the speed of development or increase bureaucracy.
- **Activity:** This involves doing the actual writing of code as opposed to all the planning that sometimes takes most of the time in software development.
- **Dexterity in motion:** This means that there must be an abundance of skills in the activity of developing code. The skills referred to are the mental skills that will arm the developers for programming challenges and team dynamics.
- **Adjustability:** This is two fold; firstly there must be room for change in the set of activities and technologies that constitute an agile development process, secondly the requirements, code, and the design/architecture must be allowed to change to the advantage of the customer.

According to Beck (1999), agile methodologies are a lightweight, efficient, low-risk, flexible, predictable, scientific, and fun way to develop software. These terms will be defined in this context to give a functional perspective of agile development.

- **Lightweight** implies minimizing everything that has to be done in the development process (e.g., documentation, requirements, etc.) in order to increase the speed and efficiency in development. The idea of minimizing documentation is still a controversial one as some assume agility to mean no documentation at all. Such views are however not unfounded because some agile extremists have expressed connotations of zero documentation claiming that the code is sufficient documentation. As agile methodologies approach higher levels of maturity minimizing documentation has evolved to generally imply providing as much documentation as the customer is willing pay for in terms of time and money.
- **Efficient** means doing only that work that will deliver the desired product with as little overhead as practically possible.
- **Low-risk** implies trading on the practical lines and leaving the unknown until it is known. In actual fact, all software development methodologies are designed to reduce the risks of project failure. At times, a lot of effort is wasted in speculative abstraction of the problem space in a bid to manage risk.
- **Predictable** implies that agile methodologies are based on what practitioners do all the time, in other words the world of ambiguity is reduced. This however does not mean that planning, designs, and architecture of software are predictable. It means that agility allows development of software in the most natural ways that trained developers can determine in advance based on special knowledge.
- **Scientific** means that the agile software development methodologies are based on sound and proven scientific principles. It nevertheless remains the responsibility of the academia to continue gathering empirical evidence on agile processes because most of the practitioners who authored agile methodologies seem to have little interest and time to carryout this kind of research.
- **Fun way** because at last developers are allowed to do what they like most (i.e., to spend most of their time writing good code that works). To the developers, agility provides a form of freedom to be creative and innovative without making the customer pay for it, instead the customer benefits from it.

Schuh (2004) defines agile development as a counter movement to 30 years of increasingly heavy-handed processes meant to refashion computer programming into software engineering, rendering it as manageable and predictable as any other engineering discipline.

On a practical perspective, agile methodologies emerged from a common discovery among practitioners that their practice had slowly drifted away from the traditional heavy document and process centered development approaches to more people-centered and less document-driven approaches (Boehm & Turner, 2004; Highsmith, 2002a; Fowler, 2002). There is a general misconception that there is no planning or there is little planning in agile processes. This is due to the fact that the agile manifesto lists as one of its four values the preference for responding to change over following a plan (Agile Alliance, 2001). In fact, planning in agile projects could be more precise than in traditional processes it is done rigorously for each increment and from a project planning perspective agile methodologies provide a risk mitigation approach where the most important principle of agile planning is feedback. Collins-Cope (2002) lists the potential risks as: risks of misunderstandings in functional requirements, risks of a deeply flawed architecture; risks of an unacceptable user interface; risks of wrong analysis and design models; risks of the team not understanding the chosen technology et cetera. Feedback is obtained by creating a working version of the system at regular intervals or per increment according to the earlier planning effort (Collins-Cope, 2002).

Besides dealing with the most pertinent risks of software development through incremental development, agile methodologies attack the premise that plans, designs, architectures, and requirements are predictable and can therefore be stabilized. Agile methodologies also attack the premise that processes are repeatable (Highsmith, 2001; Schwaber & Beedle, 2002). These two premises are part of fundamental principles on which traditional methodologies are built, and they also happen to be the main limitations of the traditional methodologies.

Boehm et al. (2004) view agile methodologies as a challenge to the mainstream software development community that presents a counter-culture movement, which addresses change from a radically different perspective. All agile methodologies follow the four values and 12 principles as outlined in the agile manifesto.

CONTEXTUAL DEFINITION

From these definitions of agile methodologies, a contextual definition can be derived which looks at what agility means in terms of certain specific software engineering concepts. Examples of that would be concepts are software quality assurance, software process improvement, software process modeling, and software project management. Agile methodologies will now be defined according to these concepts. Since this book is specifically focused on agile software quality assurance the definition of agile software quality assurance will be given in more detail.

AGILE SOFTWARE QUALITY ASSURANCE

This section starts by summarizing the traditional definitions of quality and then presents a summary of the work that has been done in the area of agility and quality. References to older literature on software quality are not intended to be exhaustive, but to be simply present a fare baseline for evaluating software quality perspectives in the modern processes. The authors are aware of a number of initiatives in research and academic institutions where evaluation of quality concepts is performed on some agile practices.

DEFINING QUALITY

Have you ever wondered what Joseph Juran generally considered to be a quality legend would have said about agile processes and the quality movement? Well, this is what he said about the ISO 9000 when he was asked by Quality Digest if he thought ISO 9000 had actually hindered the quality movement; "Of course it has. Instead of going after improvement at a revolutionary rate, people were stampeded into going after ISO 9000, and they locked themselves into a mediocre standard. A lot of damage was, and is, being done" (QCI International, 2002).

According to Juran, quality is fitness for use, which means the following two things: "(1) quality consists of those product features that meet the needs of the customers and thereby provide product satisfaction. (2) Quality consists of freedom from deficiencies" (Juran & Gryna, 1988).

Philip Crosby, who developed and taught concepts of quality management, whose influence can be found in the ISO 9000:2000 standard, which differs from the 1994 standard in the context of each of the eight principles, defines quality as conformance to requirements and zero defects (Crosby, 1984).

ISO 9000 defines quality as the totality of characteristics of an entity that bear on its ability to satisfy stated or implied needs. Where "stated needs" means those needs that are specified as requirements by the customer in a contract, and 'implied needs' are those needs that are identified and defined by the company providing the product. These definitions of quality have a general bias towards the manufacturing industry although they should in general apply to all products, nevertheless, software products are rather complex hence they should be defined in a slightly different way.

Weinberg defines quality simply as "the value to some people" (Weinberg, 1991) and some have expanded on that to mean the association of quality with human assessment, and cost and benefit (Hendrickson, 2004).

Some software engineers have defined software quality as follows:

1. Meyer (2000) defines software quality according to an adapted number of quality parameters as defined by McCall (1977), which are correctness, robustness, extendibility, reusability, compatibility, efficiency, portability, integrity, verifiability, and ease of use.
2. Pressman, who derives his definition from Crosby, defines quality as a "conformance to explicitly stated functional requirements, explicitly documented development standards, and implicit characteristics that are expected of all professionally developed software" (Pressman, 2001).
3. Sommerville (2004) defines software quality as a management process concerned with ensuring that software has a low number of defects and that it reaches the required standards of maintainability, reliability, portability, and so on.
4. van Vliet (2003) follows the IEEE definition of quality as stated in the IEEE Glossary of Software Engineering Terminology, which defines quality assurance in two ways as: "(1) A planned and systematic pattern of all actions necessary to provide adequate confidence that the item or product conforms to established operational, functional, and technical requirements. (2) A set of activities designed to evaluate the process by which products are developed or manufactured" (IEEE, 1990). van Vliet's perspective then combines this definition with the analysis of the different taxonomies on quality.
5. Pfleeger (2001) aligns her perspective with Garvin's quality perspective, which views quality from five different perspectives namely; the transcendental meaning that quality can be recognized but not defined,

user view meaning that quality is fitness for purpose, manufacturing meaning that quality is conformance to specification, product view meaning that quality is tied to inherent product characteristics, and the value-based view meaning that quality depends on the amount the customer is willing to pay for the product.

6. Bass (2006) argues that the common practice of defining software quality by dividing it into the ISO 9126 (i.e., functionality, reliability usability, efficiency maintainability, and portability) does not work. His argument is that "in order to use a taxonomy, a specific requirement must be put into a category" (Bass, 2006). However, there are some requirements that may be difficulty to put under any category, for example, "denial of service attack, response time for user request, etc." What Bass (2006) then

Table 1. Agile quality techniques as applied in extreme programming

Technique	Description
Refactoring	Make small changes to code, Code behaviour must not be affected, Resulting code is of higher quality (Ambler, 2005).
Test-driven development	Create a test, Run the test, Make changes until the test passes (Ambler, 2005).
Acceptance testing	Quality assurance test done on a finished system, Usually involves the users, sponsors, customer, etc. (Huo, Verner, Zhu, & Babar, 2004).
Continuous integration	Done on a daily basis after developing a number of user stories. Implemented requirements are integrated and tested to verify them. This is an important quality feature.
Pair programming	Two developers work together in turns on one PC, Bugs are identified as they occur, Hence the product is of a higher quality (Huo et al., 2004).
Face-to-face communication	Preferred way of exchanging information about a project as opposed to use of telephone, email, etc. Implemented in the form of daily stand-up meetings of not more than twenty minutes (Huo et al, 2004). This is similar to the daily Scrum in the Scrum method. It brings accountability to the work in progress, which vital for quality assurance.
On-site customer	A customer who is a member of the development team, Responsible for clarifying requirements (Huo et al., 2004).
Frequent customer feedback	Each time there is a release the customer gives feedback on the system, and result is to improve the system to be more relevant to needs of the customer (Huo et al., 2004). Quality is in fact meeting customer requirements.
System metaphor	Simple story of how the system works (Huo et al., 2004), Simplifies the discussion about the system between customer/ stakeholder/ user and the developer into a non-technical format. Simplicity is key to quality.

proposes is the use of quality attributing general scenarios.

From an agile perspective, quality has been defined by some practitioners as follows:

McBreen (2003) defines agile quality assurance as the development of software that can respond to change, as the customer requires it to change. This implies that the frequent delivery of tested, working, and customer-approved software at the end of each iteration is an important aspect of agile quality assurance.

Ambler (2005) considers agile quality to be a result of practices such as effective collaborative work, incremental development, and iterative development as implemented through techniques such as refactoring, test-driven development, modelling, and effective communication techniques.

To conclude this section, Table 1 gives a summary of the parameters that define agile quality as specifically applied in extreme programming--a popularly used agile methodology. These aspects of agile quality have eliminated the need for heavy documentation that is prescribed in traditional processes as a requirement for quality. Quality is a rather abstract concept that is difficult to define but where it exists, it can be recognized. In view of Garvin's quality perspective there may be some who have used agile methodologies in their software development practices and seen improvement in quality of the software product but could still find it difficult to define quality in the agile world.

EVALUATING QUALITY IN AGILE PROCESSES

So can we evaluate quality assurance in agile processes? This can be done through:

- The provision of detailed knowledge about specific quality issues of the agile processes.
- Identification of innovative ways to improve agile quality.
- Identification of specific agile quality techniques for particular agile methodologies.

Literature shows that Huo et al. (2004) developed a comparison technique whose aim was to provide a comparative analysis between quality in the waterfall development model (as a representative of the traditional camp) and quality in the agile group of methodologies. The results of the analysis showed that there is indeed quality assurance in agile development, but it is achieved in a different way from the traditional processes. The limitations of Huo et al.'s tool however, are that the analysis:

- Singles out two main aspects of quality management namely quality assurance and verification and validation.
- Overlooks other vital techniques used in agile processes to achieve higher quality management.
- Agile quality assurance takes quality issues a step beyond the traditional software quality assurance approaches.

Another challenge of Huo et al.'s technique is that while the main purpose of that analysis was to show that there is quality assurance in agile processes, it does not make it clear what the way forward is. Agile proponents do not seem to be worried about comparison between agile and traditional processes as some of the more zealous "agilists" believe that there is no way traditional methods can match agile methods in any situation (Tom Poppendieck, personal e-mail 2005).

The evaluation described in this section improves on (Huo et al., 2004) framework by further

Table 2. Mapping software quality parameters to agile techniques

Software Quality Parameters	Agile Techniques	Possible Improvements
Correctness	Write code from minimal requirements. Specification is obtained by direct communication with the customer. Customer is allowed to change specification. Test-driven development.	Consider the possibility of using formal specification in agile development, Possible use of general scenarios to define requirements (note that some development teams are already using this).
Robustness	Not directly addressed in agile development.	Include possible extreme conditions in requirements.
Extendibility	A general feature of all OO developed applications. Emphasis is on technical excellence and good design. Emphasis also on achieving best architecture.	Use of modeling techniques for software architecture.
Reusability	A general feature of all OO developed applications. There are some arguments against reusability of agile products (Turk, France, & Rumpe, 2002; Weisert, 2002).	Develop patterns for agile applications.
Compatibility	A general feature of all OO developed applications.	Can extra features be added for the sake of compatibility even if they may not be needed? This could contradict the principle of simplicity.
Efficiency	Apply good coding standards.	Encourage designs based on the most efficient algorithms
Portability	Practice of continuous integration in extreme programming.	Some agile methods do not directly address issues of product deployment. Solving this could be to the advantage of agility.
Timeliness	Strongest point of agility, Short cycles, quick delivery, etc.	
Integrity	Not directly addressed in agile development.	
Verifiability	Test-driven development is another strength of agility.	
Ease of use	Since the customer is part of the team, and customers give feedback frequently, they will most likely recommend a system that is easy to use.	Design for the least qualified user in the organization.

identifying some more agile quality techniques and then in an innovative way identifies the agile process practices that correspond to each technique. The contribution of this evaluation is the further identification of possible practices that can be done to improve on the already high quality achievements enjoyed by agile processes.

TECHNIQUES

The parameters that define software quality from a top-level view can be rather abstract. However, the proposed technique picks each of the parameters and identifies the corresponding agile techniques that implement the parameter in one way or another. Possible improvements to the current practice have been proposed by analysing the way agile practitioners work. Of great importance to this kind of analysis is a review of some of the intuitive practices that developers usually apply which may not be documented. You may wonder how much objectivity can be in such information. The point though is that developers tell their success stories at different professional forums and some of the hints from such deliberations have been captured in this technique without following any formal data gathering methodology. The authors believe that gathering of informal raw data balances the facts especially in cases where developers talk about their practice. Once the data is gathered formally, then a lot of prejudices and biases come in and there will be need to apply other research techniques to balance the facts. Tables 2 and 3 summarize the evaluation approach.

In formal software quality management, quality assurance activities are fulfilled by ensuring that each of the parameters listed in Table 2 are met to a certain extent in the software development life cycle of the process concerned. A brief definition of each of these parameters is given according to Meyer (2000):

- **Correctness:** The ability of a system to perform according to defined specification.
- **Robustness:** Appropriate performance of a system under extreme conditions. This is complementary to correctness.
- **Extendibility:** A system that is easy to adapt to new specification.
- **Reusability:** Software that is composed of elements that can be used to construct different applications.
- **Compatibility:** Software that is composed of elements that can easily combine with other elements.
- **Efficiency:** The ability of a system to place as few demands as possible to hardware resources, such as memory, bandwidth used in communication and processor time.
- **Portability:** The ease of installing the software product on different hardware and software platforms.
- **Timeliness:** Releasing the software before or exactly when it is needed by the users.
- **Integrity:** How well the software protects its programs and data against unauthorized access.
- **Verifiability:** How easy it is to test the system.
- **Ease of use:** The ease with which people of various backgrounds can learn and use the software.

SOFTWARE PROCESS IMPROVEMENT

A bigger-picture view of agile processes leads to a notion that agile methods are a group of processes that have reduced the development timeframe of software systems and introduced innovative techniques for embracing rapidly changing business requirements. With time, these relatively new techniques should develop into mature software engineering standards.

SOFTWARE PROCESS MODELING

The agile perspective to software process modeling is that whether formal or informal when approaches to modeling are used the idea is to apply modeling techniques in such a way that documentation is minimized and simplicity of the desired system is a virtue. Modeling the agile way has led to breakthroughs in the application of agile methods to the development of large systems (Ambler, 2002)

SOFTWARE PROJECT MANAGEMENT

The agile approach to managing software projects is based on giving more value to the developers than to the process. This means that management should strive to make the development environment conducive. Instead of worrying about critical path calculation and Gantt chart schedules, the project manager must facilitate face-to-face communication, and simpler ways of getting feedback about the progress of the project. In agile development there is need to be optimistic about people and assume that they mean good hence give them space to work out the best way to accomplish their tasks. It is also an agile strategy to trust that people will make correct professional decisions about their work and to ensure that the customer is represented in the team throughout the project.

THE AGILE METHODOLOGY EVALUATION FRAMEWORK

All agile methodologies have striking similarities amongst their processes because they are based on the four agile values and 12 principles. It is interesting to note that even the authors of agile methodologies no longer emphasize their methodology boundaries and would use practices from other agile methodologies as long they suit a given situation (Beck & Andres, 2004). In fact, Kent Beck in his extreme programming (XP) master classes frequently mentions the errors of extremism in the first edition of his book on XP (Beck, 1999). A detailed review of agile methodologies reveals that agile processes address the same issues using different real life models.

The evaluation technique presented in this chapter reveals, for example, that lean development (LD) views software development using a manufacturing and product development metaphor. Scrum views software development processes using a control engineering metaphor. Extreme programming views software development activities as a social activity where developers sit together. Adaptive systems development (ASD) views software development projects from the perspective of the theory of complex self-adaptive systems (Mnkandla, 2006).

Tables 3 to 6 summarize the analysis of agile methodologies. Only a few of the existing agile methodologies have been selected to illustrate the evaluation technique. The first column from the left on Tables 3, 4, and 5 lists some methodology elements that have been chosen to represent the details of a methodology. There is a lot of subjectivity surrounding the choice of methodology elements. It is not within the scope of this chapter to present a complete taxonomy of methodologies. For more detailed taxonomies see Avison and Fitzgerald (2003), Boehm et al. (2004), Glass and Vessey (1995), and Mnkandla (2006). Therefore, the elements used here were chosen to reveal the similarities amongst different agile methodologies. The importance of revealing these similarities is to arm the developers caught up in the agile methodology jungle wondering which methodology to choose. While the methodology used in your software development project may not directly lead to the success of a project and may not result in the production of a high quality product use of a wrong methodology will lead to project failure. Hence, there is in wisdom selecting a correct and relevant process. Most organization may not afford the luxury of using different methodologies for each project though that would be ideal for greater achievements. It also sounds impractical to have a workforce that is proficient in many methodologies. Sticking to one methodology and expect it to be sufficient for all projects would also be naïve (Cockburn, 2000). This evaluation technique therefore gives software development organizations an innovative wit to tailor their development process according to the common practices among different agile methodologies. The advantage is to use many methodologies without the associated expenses of acquiring them.

There is a need to understand in detail each agile methodology that will be analyzed so as to reveal the underlying principles of the methodology. This technique gives the background details as to why the methodology was developed in the first place. An answer to this question would reveal the fundamental areas of concern of the methodology and what fears the methodology addresses. The prospective user of the methodology would then decide whether such concern area is relevant to their project. Identifying what problems the methodology intends to solve is another concern of this evaluation. Some methodologies have a general bias toward solving technical problems within the development process (i.e., extreme programming deals with issues such as how and when to test the code). There are other agile methodologies that solve project management problems (i.e., Scrum deals with issues such as how to effectively communicate within a project). Yet other agile methodologies solve general agile philosophy problems (i.e., Crystal deals with issues such as the size of the methodology vs. the size of the team and the criticality of the project. There may be other agile methodologies that solve a mix of problems right across the different categories mentioned here for example Catalyst puts some project management aspects into XP (see www.ccpace.com for details on Catalyst).

Evaluation of each methodology should also reveal what sort of activities and practices are prevalent in the methodology. This should assist prospective users of the methodology to determine the practices that could be relevant to their given situation. This evaluation technique reveals that some of the practices from different methodologies actually fulfill the same agile principles and it would be up to the developers to decide which practices are feasible in their situation. Therefore, the implication is that at the level of implementation it becomes irrelevant which agile methodology is used, for more on this concept see Mnkandla (2006). Another aspect of agile methodologies revealed by this evaluation technique is what the methodology delivers at the end of the project. When a developer looks for a methodology, they usually have certain expectations about what they want as an output from the methodology. Hence, if the methodology's output is not clearly understood problems may result. For example if the developer expects use of the methodology to lead to the delivery of code and yet the aim of the methodology is in fact to produce a set of design artifacts such as those delivered by agile modeling this could lead to some problems. Finally, this evaluation technique also reveals the domain knowledge of the author of the methodology. In this phase of analysis, there is no need to mention any names of the authors but simply to state their domain expertise. The benefit of revealing the background of the methodology author is to clarify the practical bias of the methodology, which is usually based on the experience, and possible fears of the methodology's author.

Tables 3 to 5 give a summary of the analysis of specific agile methodologies to illustrate how this analysis technique can be used for any given agile methodologies.

Table 3. Analyzing scrum methodology

Elements	Description
Real Life Metaphor	Control engineering.
Focus	Management of the development process.
Scope	Teams of less than 10, but is scalable to larger teams.
Process	Phase 1: planning, product backlog, & design. Phase 2: sprint backlog, sprint. Phase 3: system testing, integration, documentation, and release.
Outputs	Working system.
Techniques	Sprint, scrum backlogging (writing use cases).
Methodology Author (two)	1. Software developer, product manager, and industry consultant. 2. Developed mobile applications on an open technology platform. Component technology developer. Architect of advanced internet workflow systems.

ANALYZING SCRUM

Scrum has been in use for a relatively longer period than other agile methodologies. Scrum, along with XP, is one of the more widely used agile methodologies. Scrum's focus is on the fact that "defined and repeatable processes only work for tackling defined and repeatable problems with defined and repeatable people in defined and repeatable environments" (Fowler, 2000), which is obviously not possible. To solve the problem of defined and repeatable processes, Scrum divides a project into iterations (which are called Sprints) of 30 days. Before a Sprint begins, the functionality required is defined for that Sprint and the team is left to deliver it. The point is to stabilize the requirements during the Sprint. Scrum emphasizes project management concepts (Mnkandla & Dwolatzky, 2004b) though some may argue that Scrum is as technical as XP. The term Scrum is borrowed from Rugby: "A Scrum occurs when players from each team clump closely together…in an attempt to advance down the playing field" (Highsmith, 2002b). Table 3 shows application of the analysis technique to Scrum.

ANALYZING LEAN DEVELOPMENT

Lean software development like dynamic systems development method and Scrum is more a set of project management practices than a definite process. It was developed by Bob Charette and it draws on the success that lean manufacturing gained in the automotive industry in the 1980s. While other agile methodologies look to change the development process, Charette believes that to be truly agile, there is need to change how companies work from the top down (Mnkandla et al., 2004b). Lean development is targeted at changing the way CEOs consider change with regards to management of projects. LD is based on lean thinking whose origins are found in lean production started by Toyota Automotive manufacturing company (Poppendeick & Poppendeick, 2003). Table 4 shows application of the analysis technique to LD.

Table 4. Analyzing lean development methodology

Elements	Description
Real Life Metaphor	Manufacturing and product development.
Focus	Change management. Project management.
Scope	No specific team size.
Process	Has no process.
Outputs	Provides knowledge for managing projects.
Techniques and Tools	Lean manufacturing techniques.
Methodology Author	Research engineer at the US Naval Underwater Systems Center, author of software engineering books and papers, advisory board in project management.

ANALYZING EXTREME PROGRAMMING

Extreme programming (XP) is a lightweight methodology for small-to-medium-sized teams developing software based on vague or rapidly changing requirements (Beck, 1999). In the second version of XP, Beck extended the definition of XP to include team size and software constraints as follows (Beck et al., 2004):

- **XP is lightweight:** You only do what you need to do to create value for the customer.
- **XP adapts to vague and rapidly changing requirements:** Experience has shown that XP can be successfully used even for project with stable requirements.
- **XP addresses software development constraints:** It does not directly deal with project portfolio management, project financial issues, operations, marketing, or sales.
- **XP can work with teams of any size:** There is empirical evidence that XP can scale to large teams.

Software development using XP starts from the creation of stories by the customer to describe the functionality of the software. These stories are small units of functionality taking about a week or two to code and test. Programmers provide estimates for the stories, the customer decides, based on value and cost, which stories to do first. Development is done iteratively and incrementally. Each two weeks, the programming team delivers working stories to the customer. Then the customer chooses another two weeks worth of work. The system grows in functionality, piece by piece, steered by the customer. Table 5 shows application of the analysis technique to XP.

A WALK THROUGH THE ANALYSIS TECHNIQUE

Each of the methodology elements as represented in Tables 3 to 5 will be defined in the context of this analysis approach.

METHODOLOGY'S REAL LIFE METAPHOR

This element refers to the fundamental model/ metaphor and circumstances that sparked the initial idea of the methodology. For example watching the process followed by ants to build an anthill could spark an idea of applying the same process to software development.

Table 5. Analyzing extreme programming

Elements	Description
Real Life Metaphor	Social activity where developers sit together.
Focus	Technical aspects of software development.
Scope	Less than ten developers in a room. Scalable to larger teams.
Process	**Phase 1:** Writing user stories. **Phase 2:** Effort estimation, story prioritization. **Phase 3:** Coding, testing, integration testing. **Phase 4:** Small release. **Phase 5:** Updated release. **Phase 6:** Final release (Abrahamsson et al, 2002).
Outputs	Working system.
Techniques and Tools	Pair programming, refactoring, test-driven development, continuous integration, system metaphor.
Methodology Authors (two)	1. Software developer (Smalltalk). Strong believer of communication, reflection, and innovation. Pattern for software. Test-first development. 2. Software developer (Smalltalk). Director of research and development. Developed the Wiki. Developed Framework for Integrated Test (Fit).

METHODOLOGY FOCUS

The focus of the methodology refers to the specific aspects of the software development process targeted by the methodology. For example, agile modeling targets the design aspects of the software development process and also considers issues of how to model large and complex projects the agile way.

METHODOLOGY SCOPE

This element outlines the details to which the methodology's development framework is spelled out. This is where the methodology specifies what it covers within a project. The importance of this parameter is to help the user to identify the list of tasks that the methodology will help manage. Remember a methodology does not do everything but simply gives guidelines that help in the management of a project. The scope of a software development project is relevant in determining the size of the team.

METHODOLOGY PROCESS

This parameter describes how the methodology models reality. The model may be reflected in the life cycle or development process of the methodology. The model provides a means of communications, captures the essence of a problem or a design, and gives insight into the problem area (Avison et al., 2003). The importance of this parameter is that it gives the user a real worldview of the series of activities that are carried out in the development process.

METHODOLOGY OUTPUTS

This parameter defines the form of deliverables to be expected from the methodology. For example if an organization purchased lean development methodology today, would they get code from application of the methodology, or would they get some documents, etc (Avison et al., 2003). Each agile methodology will give different outputs hence the user can choose the methodology that gives them the output they require.

TECHNIQUES AND TOOLS

This parameter helps the user to identify the techniques and tools applicable to the methodology. Tools may be software applications that can be used to automate some tasks in the development process, or they can be as simple as whiteboards and flip charts. In fact, it is the use of tools that makes the implementation of a methodology enjoyable. Organizations therefore tend to spend a lot of money acquiring tools and training staff on tools. As technology evolves and new tools emerge, more acquisitions and training are usually done. However, most agile methodologies do not specify tools and most agile practitioners use open source tools, which reduces potential costs on software tools.

Each methodology has its own techniques that may be relevant or irrelevant to the problem at hand. Examples of techniques in extreme programming would be pair programming, and the scrum meeting in Scrum methodology. The user then analyzes these techniques in relation to the present project, to determine need for the techniques and include variations that will be part of tailoring the methodology.

METHODOLOGY AUTHOR

This parameter defines the domain knowledge of the methodology author. The benefit of doing this is to clarify the background from which the methodology was conceived. There is no need to mention the name of the author or a detailed biography of the methodology author.

Table 6 summarizes the phase of the analysis where all the practices are brought together and similar practices are identified across different methodologies.

Table 6 classifies the practices using the superscripts 1, 2, 3, 4, and 5. The practices with

Table 6. Identifying similarities among the practices

	Practices
XP	The planning process[1], small releases[2], metaphor, test-driven development[2], story prioritization[3], collective ownership[3], pair programming[3], forty-hour work week[3], on-site customer[4], refactoring[5], simple design[5], and continuous integration[5].
LD	Eliminate waste[1], minimize inventory[1], maximize flow[2], pull from demand[2], meet customer requirements[2], ban local optimization[2], empower workers[3], do it right the first time[4], partner with suppliers[4], and create a culture of continuous improvement[5].
Scrum	Capture requirements as a product backlog[1], thirty-day Sprint with no changes during a Sprint[2], Scrum meeting[3], self-organizing teams[3], and Sprint planning meeting[4].

the same superscript implement the same agile principle.

- "1" represents practices that deal with planning issues such as requirements gathering. The three methods shown here use different terms but the principle is to capture minimal requirements in the simplest available way and start coding.
- "2" represents practices that deal with improvement of quality in terms of meeting the volatile requirements.
- "3" represents practices that facilitate freely working together of developers, effective communication, empowered decision-making, and team dynamics issues.
- "4" represents practices that deal with quick delivery of the product.
- "5" represents practices that deal with agile quality assurance property of ensuring that the product is improved continuously until deployment.

When the similar practices are identified, the developers can then decide to select and tailor some practices to their environment according to relevance, and project and customer priorities. You will notice that the choice of the activities of the development process according to this analysis have shifted from focusing on the individual methodologies to a focus on practices.

ISSUES AND CONTROVERSIES SURROUNDING AGILE DEVELOPMENT

Software Development Common Ground

This section looks at issues that are done in a similar way among different software development methodologies. Most software development processes in use today involve some of the following activities: planning, estimation, and scheduling of the tasks, design, coding, and testing, and deployment and maintenance. What varies among the different processes is the sequence followed in implementing each of the phases, and the level of detail to which each phase is carried out. Some methodologies may implement all of the activities and some partial methodologies may specialize in just a few. The other difference is in the way the process values the people involved in the development activities and what value is attached to the customer in relation to what needs to be done. These differences mark the major boundaries among software development methodologies.

Agile Development Higher Ground

This section looks at issues that are done in a peculiar way by agile methodologies. The role of the domain expert in agile methodologies is rather unique. Software development experts with practical experience in this field have a lot of knowledge that can be classified as *tacit knowledge* due to the fact that it is gained through practice and is not written down in any form. Tacit knowledge is difficult to quantify hence this concept remains quite subjective in the implementation of agile methodologies. However, the strength of using tacit knowledge rests in the team spirit that puts trust on experts to do what they know best within their professional ethics. This in fact is what differentiates the "agile movement" from other development processes. Another hot issue about agile development is the concept of *self-organizing* teams. This concept means that agile development teams are allowed to organize themselves the best way they want in order to achieve the given goals. As a result of applying this concept, managing agile projects becomes different from the traditional approaches to project management. The role of a project manager becomes more of a facilitator than a controller. Detailed discussions on what has become known as "agile project management" can be found in Highsmith (2004) and Schwaber (2004) and at http://finance.groups.yahoo.com/group/agileprojectmanagement/.

Agile methodologies also emphasize on *light documentation*. This concept has been quite controversial since agile methodologies started. The main reason for the controversy is that the traditional methodologies have always associated documentation with proper planning, software quality assurance, deployment, user training, maintenance, etc. Agile methodologists however, believe that documentation should be minimum because of the associated expenses. In agile methodologies, the general belief is that correctly written code is sufficient for maintenance. The *Test first* technique, which was originally an XP practice and is now widely applied in other agile methodologies is another peculiar agile practice (though its origins may be from earlier processes). The test first technique is a software development approach that implements software design through writing tests for each story before the code is written. The test code then amounts to design artifacts and replaces the need for design diagrams etc.

Challenges Faced by Agile Development

This section looks at issues that are still grey areas to agile methodologies. One of the interesting issues about agility is what is going to happen to the issues of innovative thinking embedded in agile development as the processes attain higher and higher levels of maturity and quality assurance. Are we going to see a situation where agility retires and fails to be agile? Another area of software development that is always heavily debated at agile gatherings is the area of how to cost projects that are developed the agile way. The main difficulty is estimating the cost of an entire project based on iterations. There has been some effort towards dealing with this challenge especial at big agile conferences, for example the extreme programming and agile processes in software engineering held in Europe once per year (see www.XP2006.org). Another example is the Agile Development Conference also held once per year in the USA (see www. agiledevelopmentconference.com).

As agile processes begin to enter grounds such as enterprise architecture, patterns, and software reuse, their software process jacket is getting heavier and heavier and if this is not watched by agile proponents we might have a situation sometime in the future where another software development revolution emerges to maintain the legacy of agility.

THE FUTURE TRENDS OF AGILE SOFTWARE DEVELOPMENT

Agile methodologies are certainly moving toward higher levels of maturity due to a number of things. The first contribution to agile maturity is the availability of comprehensive sources of simple descriptive and analytical information about agile methodologies. The second contribution to agile maturity is the growth in academic research interest in agility, which has resulted in a lot of empirical data being collected and scientifically analyzed to prove and disprove anecdotal data about agile processes. The third contribution to agile maturity is the massive exchange of practical experiences amongst the different practitioners involved in agile software development. The general direction of the agile movement seems to be towards more and more demand for the adoption of agile practices by the larger organizations that have been traditionally associated with traditional processes. There have been reports of higher demands for agile consultancy and training as more and more organizations adopt agile development practices, Poppendieck (personal communication, November 07, 2005) said there was more demand in North America, Europe, and even Japan where his book on lean software development sold more than ten thousand copies. Another interesting development by the agile alliance is their offer to sponsor agile research. This will certainly go a long way in boosting the process maturity of agile methodologies.

CONCLUSION

In this chapter, an overview of agile methodologies was presented without going into the details of describing each existing agile methodology. The focus of the chapter was to provide an informed review of agile methodologies that included a comprehensive definition of what agility and agile quality assurance is all about. The approach to the definition as presented in this chapter was to give a theoretical definition, which is the perspective of those who are philosophical about agile methodologies, a practical definition, which is the perspective of those who are on the development work floors, and a contextual definition, which is a perspective based on the different contexts of the activities of the software development process. In order to enhance understanding of the agile processes, an analysis model was presented. The philosophy of this technique is to cut deep into each given agile methodology and reveal the core values, principles, and practices of the methodology so as to compare the common activities among different agile processes. The aim of doing such an analysis is to provide a technique for striking the balance between these two extremes: "getting lost in the agile methodology jungle and holding onto one methodology." The benefit of using this analysis method is the attainment of a deeper understanding of all the agile methodologies analyzed. This should lay the ground for training and adoption of agile methodologies from a generic point of view rather than worrying about individual agile methodologies.

REFERENCES

Abrahamsson, P., Salo, O., Ronkainen, J., & Warsta, J. (2002). *Agile software development methods: Review and analysis. VVT Publications*, (478), 7-94.

Agile Alliance. (2001). *Manifesto for agile software development.* Retrieved May 2, 2005, from http://www.agilemanifesto.org

Ambler, S. W. (2002). *Agile modeling.* John Wiley and Sons.

Ambler, S. W. (2003). *Agile database techniques: Effective strategies for the agile software developer* (pp. 3-18). John Wiley & Sons.

Ambler, S. (2005). *Quality in an agile world. Software Quality Professional*, 7(4), 34-40.

Avison, D. E., & Fitzgerald, G. (2003). *Information systems development: Methodologies techniques and tools.* McGraw-Hill.

Bass, L. (2006, January). Designing software architecture to achieve quality attribute requirements. *Proceedings of the 3rd IFIP Summer School on Software Technology and Engineering* (pp. 1-29). South Africa.

Beck, K. (1999). *Extreme programming explained: Embrace change* (pp. 10-70). Reading, MA: Addison-Wesley.

Beck, K., & Andres, C. (2004). *Extreme programming explained: Embrace change.* Addison-Wesley Professional.

Boehm, B., & Turner, R. (2004). *Balancing agility and discipline: A guide for the perplexed* (1st ed., pp. 165-194, Appendix A). Addison-Wesley.

Brandt, I. (1983). A comparative study of information systems development methodologies, Proceedings of the IFIP WG8.1 Working Conference on Feature Analysis of Information Systems Design Methodologies. In T. W. Olle, H. G. Sol,

& C. J. Tully (Eds.), *Information systems design methodologies: A feature analysis* (pp. 9-36). Amsterdam: Elsevier.

Cockburn, A. (2000). Selecting a project's methodology. *IEEE Software*, 64-71.

Cohen, D., Lindvall, M., & Costa, P. (2003). *Agile software development* (pp. 11-52). Fraunhofer Center for Experimental Software Engineering, Maryland, DACS SOAR 11 Draft Version.

Collins-Cope, M. (2002). *Planning to be agile? A discussion of how to plan agile, iterative, and incremental developments.* Ratio Technical Library White paper. Retrieved January 20 from http://www.ration.co.uk/whitepaper_12.pdf

Fowler, M. (2000). Put your process on a diet. *Software Development*, *8*(12), 32-36.

Fowler, M. (2002). *The agile manifesto: Where it came from and where it may go.* Martin Fowler article. Retrieved January 26, 2006, from http://martinfowler.com/articles/agileStory.html

Glass, R. L., & Vessey, I. (1995). Contemporary application domain taxonomies. *IEEE Software*, 63-76.

Hendrickson, E. (2004). *Redefining quality,* Retrieved January 12, 2006, from http://www.stickyminds.com/sitewide.asp?Function=edetail&ObjectType=COL&ObjectId=7109

Highsmith, J. (2001). The great methodologies debate: Part 1: Today, a new debate rages: Agile software development vs. rigours software development. *Cutter IT Journal*, *14*(12), 2-4.

Highsmith, J. (2002a). *Agile software development: Why it is hot!* (pp. 1-22). Cutter Consortium white paper, Information Architects.

Highsmith, J. (2002b). *Agile software development ecosystems* (pp. 1-50). Addison-Wesley.

Highsmith, J. (2004). *Agile project management.* Addison-Wesley.

Huo, M., Verner, J., Zhu, L., & Babar, M. A. (2004). Software quality and agile methods. *Proceedings of the 28th Annual International Computer Software and Applications Conference (COMPSAC04).* IEEE Computer.

IEEE. (1990). *IEEE standard glossary of software engineering terminology.* IEEE Std 610.12.

Juran, J. M., & Gryna, F. M. (1988). *Juran's quality control handbook.* Mcgraw-Hill.

Lindvall, M., Basili, V. R., Boehm, B., Costa, P., Dangle, K., Shull, F., Tesoriero, R., Williams, L., & Zelkowitz, M. V. (2002). Empirical findings in agile methods. *Proceedings of Extreme Programming and agile Methods: XP/agile Universe* (pp. 197-207).

Marick, B. (2001). Agile methods and agile testing. *STQE Magazine*, *3*(5).

McBreen, P. (2003). *Quality assurance and testing in agile projects.* McBreen Consulting. Retrieved January 12, 2006, from http://www.mcbreen.ab.ca/talks/CAMUG.pdf

Meyer, B. (2000). *Object-oriented software construction* (pp. 4-20). Prentice Hall PTR.

Mnkandla, E., & Dwolatzky, B. (2004a). Balancing the human and the engineering factors in software development. *Proceedings of the IEEE AFRICON 2004 Conference* (pp. 1207-1210).

Mnkandla, E., & Dwolatzky, B. (2004b). A survey of agile methodologies. *Transactions of the South Africa Institute of Electrical Engineers*, *95*(4), 236-247.

Mnkandla, E. (2006). *A selection framework for agile methodology practices: A family of methodologies approach.* PhD thesis, University of the Witwatersrand, Johanneburg.

Pfleeger, S. L. (2001). *Software engineering: Theory and practice.* Prentice Hall.

Poppendeick, M., & Poppendeick, T. (2003). *Lean software development: An agile toolkit for software development managers* (pp. xxi-xxviii). Addison Wesley.

Pressman, R. S. (2001). *Software engineering a practitioner's approach.* McGraw-Hill.

QCI International. (2002). *Juran: A life of quality: An exclusive interview with a quality legend.* Quality Digest Magazine. Retrieved January 12, 2006, from http://www.qualitydigest.com/aug02/articles/01_article.shtml

Schuh, P. (2004). *Integrating agile development in the real world* (pp. 1-6). MA: Charles River Media.

Schwaber, K., & Beedle, M. (2002). *Agile software development with SCRUM* (pp. 23-30). Prentice-Hall.

Schwaber, K. (2004). *Agile project management with Scrum.* Microsoft Press.

Sol, H. G. (1983). A feature analysis of information systems design methodologies: Methodological considerations. Proceedings of the IFIP WG8.1 Working Conference on Feature Analysis of Information Systems Design Methodologies. In T. W. Olle, H. G. Sol, & C. J. Tully (Eds.), *Information systems design methodologies: A feature analysis* (pp. 1-7). Amsterdam: Elsevier.

Sommerville, I. (2004). *Software engineering.* Addison-Wesley.

Turk, D., France, R., & Rumpe, B. (2002). Limitations of agile software processes. *Proceedings of the Third International Conference on eXtreme Programming and Agile Processes in Software Engineering* (pp. 43-46).

van Vliet, H. (2003). *Software engineering: Principles and practice.* John Wiley & Sons.

Weinberg, G. M. (1991). *Quality software management* (Vol. 1), *Systems Thinking.* Dorset House.

Weisert, C. (2002). *The #1 serious flaw in extreme programming (XP).* Information Disciplines, Inc., Chicago. Retrieved January 2006, from http://www.idinews.com/Xtreme1.html

Chapter II
Agile Quality or Depth of Reasoning? Applicability vs. Suitability with Respect to Stakeholders' Needs

Eleni Berki
University of Tampere, Finland

Kerstin Siakas
Alexander Technological Educational Institution of Thessaloniki, Greece

Elli Georgiadou
University of Middlesex, UK

ABSTRACT

This chapter provides a basis for discussion and analysis of the quality characteristics of the agile information systems development (AISD) process, and exposes concerns raised regarding the applicability and suitability of agile methods in different organisational and national cultures. The arguments are derived from recent literature reviews and reports on the state-of-the-art in agile methodologies. We also reflect on our own research and experience in three European countries with different academic and work values, and information systems development (ISD) industrial practices based on diverse principles. Examining the different perceptions of quality in the agile software development paradigm by comparing and contrasting them to the quality perceptions in the established ISD methodological paradigms, we aim at: (i) exploring the relationship of traditional, agile, managed, and measured ISD processes, (ii) making the human dynamics that bear on the success and consensus acceptance of IS more explicit, and (iii) establishing a critical framework/approach for reflecting on the suitability and applicability of agile methods in the development and management of quality software systems.

INTRODUCTION

Agile methods have pervaded many application domains of software development and many claim that this has occurred because agile methods (DSDM, XP, Crystal, SCRUM, ...) advocate end-user participation and satisfaction by focusing on systems realisation, requirements change, and testing as the means to achieving a "correct" information system (Beck, 2000, 2003; Beck & Fowler, 2001). On the other hand, it is argued that the agility of these new methods might lead to more complex and not well-documented systems and to a fragmented software development process (Boehm & Turner, 2003a, 2003b; Marciniak, 1994). Our motivation to examine these arguments in this chapter derives from (a) scientific and (b) practical perspectives. Naturally, a fragmented, unpredictable, and non-measurable IS process does not add to the established scientific rules that must guide software development. In that respect, the use of computational principles combined with lightweight methods, which support continuous change, might be the answer for *agile quality*, particularly for software component-based development in a post-modern information society (Berki, Georgiadou, & Holcombe, 2004; Siakas, Balstrup, Georgiadou, & Berki, 2005b).

Considering, though, the post-technological state of the global software industry, an ISD method customisation to the needs of diverse and different organisational and national cultures points rather to further argumentation for the general *applicability* and *suitability* of the agile methods paradigm. There is an obvious need for further research in order to understand the requirements of quality and the requirements of agile quality in particular, within different cultural and social contexts and, perhaps, the need to identify controllable and uncontrollable quality factors for agile ISD (Georgiadou, Siakas, & Berki, 2003). IS quality requires knowledge of different organisational and national cultures on the methods and tools used, on the ways they are used and, most importantly, on the ways people perceive quality and quality assurance (Berki, 2006; Ross & Staples, 1995; Siakas, Berki, Georgiadou, & Sadler, 1997). Awareness and application of total quality management principles and the influence of human involvement and commitment are yet unresolved and largely non-researched issues in different cultural (organisational and national) contexts. Therefore, specialised knowledge is required in order to assess, assure, and certify the quality of agile software development.

It is, yet, questionable if risks and project failures that are frequently encountered in traditional software projects could be diminished or avoided (Siakas, Georgidaou, & Berki, 2005a) by adopting an agile approach. It is, however, argued that agile methods make the key business users a very strong partner in assuring quality (Beck, 2000, 2003; Beck et al., 2001). We presume that in a mature IS society, rather than completely leaving quality to the professionals, agile development projects will perceive the key ISD stakeholders, and end-users in particular, as co-responsible for ensuring that the application fits the purpose. At present, however, one needs to compare and contrast the agile development process (agile methods and the life cycle they support) to traditional methods. Based on the results of comparisons one could, afterwards, analyse, measure, validate, and verify the method suitability and applicability derived from the agile methodology deployment. From the software quality assurance's point of view, new or/and older software quality properties could be the key attributes for constructing process and product quality principles when both traditional and agile development methods are employed in ISD.

Throughout this chapter, we proceed to a three axis quality examination of the agile methodology paradigm, outlined as follows: The chapter firstly considers a historical perspective and a discussion over process and product quality measurements in traditional software development (method and life cycle models). In order to provide an overview of the European perspective on the software process and software product quality measurements, we present supportive data drawn from case studies and questionnaires in recent cultural factors-oriented research that was carried out in four (4) European countries, namely Finland, Denmark, the UK, and Greece (Siakas, 2002; Siakas, Berki, & Georgiadou, 2003). We subsequently comment on recent research findings regarding software process and product measurements kept during traditional ISD. There are, of course, many controllable and uncontrollable factors in measuring ISD. Considering the latter, we attempt to identify and deduce suitable process and product metrics/metametrics for an agile ISD approach.

Following from the latter and considering that there are controllable and uncontrollable factors that influence any ISD process, the second axis of the chapter's discussion considers soft (*uncontrollable* and probably *non-measurable*) issues of agile ISD. This analysis elaborates on stakeholders' participation, change management, cultural and national differences, and knowledge creation and sharing. These are significant managerial issues and in today's agile and lean management development approaches are often considered to be the cornerstones for *productivity, interactivity, communication,* and *trust* in ISD teams (Berki, Isomäki, Jäkälä, 2003; Manninen & Berki, 2004).

Finally, we examine the hard issues and quality properties (rather *controllable* and *measurable* factors) that agile methods openly claim they achieve, these being *implementability, modifiability,* and *testability.* Investigating the influence of method engineering (ME) in the agile paradigm, two, rather contradictory, observations are analysed: (i) Agile methodology tool support identifies method expressability and stakeholder communication to be problems that appear frequently in both traditional and agile development methods, while (ii) continuous change of requirements (modifiability), and possibly quality of *re-engineering* and *reverse engineering* (central issues for re-factoring), could better be facilitated with the adoption of metamodelling and method engineering rules in MetaCASE and Computer aided method engineering (CAME) environments (Berki, Lyytinen, Georgiadou, & Holcombe, 2002; Kelly et al., 1996; Tolvanen, 1998).

A HISTORICAL PERSPECTIVE FOR SOFTWARE PROCESS AND PRODUCT QUALITY IMPROVEMENT

Long before the agile paradigm, in fact since 1968, software engineers have tried to emulate traditional engineering in order to address quality problems and IS failures with varying degrees of success (Georgiadou, 2003b; Sommerville, 2001). A significant source of experience was gained from quality assurance processes practised in the manufacturing industry, such as statistical process control (SPC) (Burr & Georgiadou, 1995). The emphasis has been gradually shifted from software *product* improvement to software *process* improvement, where information systems development methods (ISDMs) have long been employed to manage the software process.

Quality Trends in Information Systems Development

In the '80s and '90s emphasis was put on software process improvement (SPI) with the appearance of quality standards and guidelines, such as ISO9001 (ISO, 2005), capability maturity model integrated (CMM/CMMI) (Paulk et al., 1993), software process improvement (SPI), SPICE / ISO-15504 (Dorling, 1993), and Bootstrap (Kuvaja, 1999). More recently the Kaizen method of incremental improvements (Vitalo, Butz, & Vitalo, 2003), six sigma (George, 2003), lean manufacturing (Poppendieck & Poppendieck, 2003), and product life cycle management (PLM) (Grieves, 2005; Saaksvuori & Immonen, 2003) have been adopted by practitioners. This signifies the recognition of the fact that life cycle costs are front-loaded (i.e., more effort, time, and costs are required at the beginning of the life cycle). Agile development advocates that it is not necessary to make all the design decisions up front. Instead, the agile process and its products are constantly being improved because the developers engage in perpetual value development through brainstorming, flexible thinking and continuously maturing and improved *commitment* and *decision-making.*

Proliferation of Methods in the Software Development Life Cycle

During the last 40 years, several ISDMs have been established (Figure 1) that have been characterised as *hard* (technically oriented), *soft* (human-centered), *hybrid* (a combination of hard and soft), and *specialised* (application-oriented) (Avison & Fitzgerald, 2003; Berki et al., 2004; Georgiadou & Milankovic-Atkinson, 1995). Hard methodologies span the range of structured, object-oriented, and formal methods. Soft methodologies are exemplified by the soft systems method (SSM) (Checkland, 1981; Checkland & Scholes, 1990) and effective technical and human interaction for computer-based systems (ETHICS) (Mumford, 1983, 2003). Figure 1 provides a historical overview of the approximate year of introduction of the main exponents of prescriptive and agile methods over the last 40 years. Acronyms can be found in Appendix A.

Agile software development methods fall within the hybrid, thus a combination of hard and soft, paradigm (Berki, 2006). Extreme programming (XP) for example builds on principles first used by Checkland (1981) and Mumford (1983) in SSM and ETHICS methodologies respectively and later on by the dynamic systems development method (DSDM) (Stapleton, 1997). In particular, the agile paradigm was built on a foundation which encapsulates user participation, time boxing, (the amount of features released in a fixed amount of time), and frequent delivery of product.

By applying one of the available methodologies to the development of software, insights were gained into the problems under consideration and, thus, software product and process quality could be addressed more systematically (Berki et al., 2002). The final software-based IS should, normally and traditionally, comply with the quality requirements of timeliness, relevance, accuracy, and cost effectiveness.

From Traditional ISD to Agile ISD: User Involvement and Problem Solving

The application of solely hard or soft methods led to IS with problematic functionality and

Figure 1. Forty years of information systems development methodologies

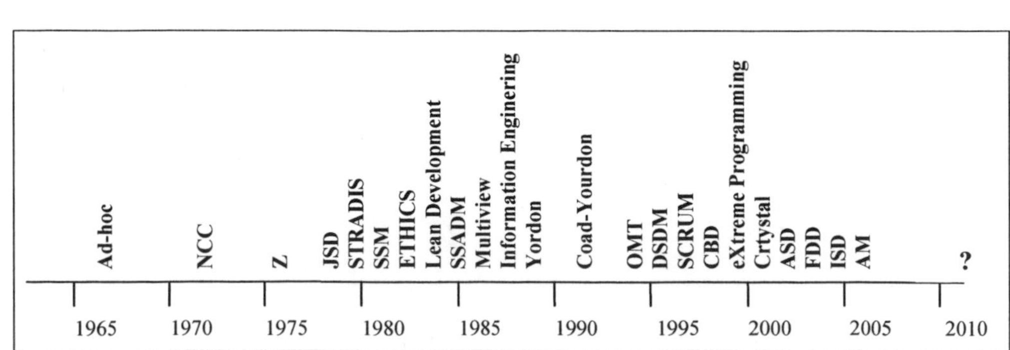

stakeholder dissatisfaction. Inevitably IS were also constructed with specialised (application-oriented) methods with no stakeholder satisfaction solutions or compliance to the most recent academic knowledge and related scientific reasoning. The appearance of hybrid methods, such as multiview (Avison & Wood Harper, 1990), indicated the need to integrate soft and hard issues. Hybrid methods concentrated on hard issues, such as implementability (Holcombe & Ipate, 1998), structuredness (Avison et al., 2003), and testability (Berki, 2004) and on soft issues, such as user participation (Mumford, 1983, 2003), conflict analysis (Checkland et al., 1990), and stakeholder communication (Berki et al., 2003). Emphasis on testing at many stages (including acceptance testing) was practised using the V and W life cycle models (Burr et al., 1995).

Traditional IS development had *some* user involvement, usually at the two ends of the development process, namely the feasibility study phase and the system acceptance phase; without, though, widely employing *formal* testing techniques (such as *acceptance* testing). The prototyping approach signalled the need for ongoing user participation. The dynamic interaction of the two constituencies, namely the users and the developers, is represented in Figure 2, which emphasises the possibility of generating many alternative solutions to a problem in a continuous interaction, trade-offs and "accommodations" throughout the process (Georgiadou et al., 1995).

Moreover, the level of user participation is inherent in each methodology. In fact, we argue that a methodology is as strong as the user involvement it supports (Berki, Georgiadou, & Siakas, 1997). Traditional participative methodologies (see SSM and ETHICS) place a strong emphasis on managerial and social issues and, usually, devote extra time and effort during the early stages of the systems life cycle. This moulds the development process and the type of the solution(s) achieved, also dependent on the lifecycle model. For instance, waterfall model emphasis, stepwise refinement, and problem reduction, while prototyping approach denotes an evolutionary and sometimes state transition approach. User involvement is mainly through discussions with the developers and participation in walkthroughs, reviews, and inspections. Larman (2004) argues that lessons learned from prototyping and iterative development as well as cyclic models, such as the spiral model, paved the way for the evolution of the agile paradigm.

Comparing the early aims of software engineering as a discipline to handle the unfore-

Figure 2. The components in the development process: The "balanced" interaction—prototyping

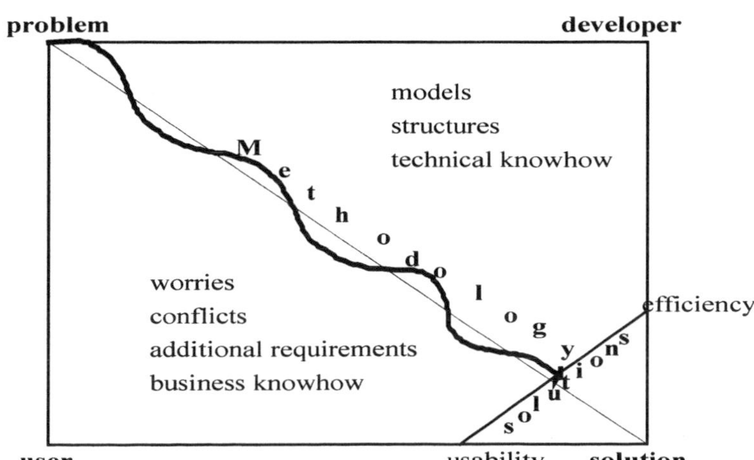

seen software crisis of the 60s by a disciplined method(ology) to the aims of the Agile Manifesto (2001), one could conclude that there is, now, a new software engineering challenge. Attributing though the appearance of the agile methods to a traditional method(ology) crisis (Berki, 2006) might lead to a deeper understanding of the reasons that agile methods have been accepted, discussed, and/or criticised in equal measures. Notwithstanding, traditional ISDMs have retained a role as quality assurance instruments for the software development process, and there has been substantial evidence in theory and practice for living up to such expectations (Avison et al., 1990; Boehm & Turner, 2003a, 2003b Jayaratna, 1994).

PROCESS AND PRODUCT QUALITY METRICS IN TRADITIONAL AND AGILE SOFTWARE DEVELOPMENT

Failure and success of IT/IS projects have been discussed since the early 70s when organisations started to use computer technology to harness the ability of their information systems (Lyytinen & Hirschheim, 1987). IT project failures have been widely reported and studied (Dalcher & Genus, 2003; Heeks, 2002; Moussa & Schware, 1992). Failures range from total malfunction to abandonment, rejection, and non-use. Statistics presented in Johnson and Foote (1988) revealed that five out of six software projects were considered unsuccessful, and approximately a third of software projects were cancelled. The remaining projects delivered software at almost double the expected budget and time to develop than originally planned.

Considerable advances have also been achieved by the introduction of methods and tools for the systematic planning and control of the development process (Avison et al., 2003; Berki, 2001; Georgiadou, 2003b; Jackson, 1994; Jayaratna, 1994). Despite these efforts, systems continue to fail with dramatic frequency (Dalcher et al., 2003). Systems development methodologies have been proposed and used to address the problems of ambiguous user requirements, non-ambitious systems design, unmet deadlines, exceeded budgets, poor quality software with numerous "bugs" and poor documentation. These and other factors made IS inflexible to future changes and difficult to maintain.

Quality and Software Process Improvement

Software process improvement (SPI) has become a practical tool for companies where the quality of the software is of high value (Järvinen, 1994). In a technical report with results from 13 organisations, and with the number of post-release defect reports used as a measure, Herbsleb, Carleton, Rozum, Siegel, and Zubrow (1994) showed that due to software process improvement (SPI), the products and business value (especially return on investment – ROI) was improved. It is generally considered that a well documented and a repeatable process is essential for developing software products of high quality. There is also evidence that the use of standards and process assessment models has a positive impact on the quality of the final software product (Kitchenham & Pfleeger, 1996).

The software engineering community has gradually moved from product quality-centred corrective methods to process quality-centred, preventive methods, thus shifting the emphasis from product quality improvement to process quality improvement. Inspections at the end of the production line have long been replaced by design walkthroughs and built-in quality assurance techniques throughout the development life cycle (Georgiadou et al., 2003). Quality is an elusive, complex concept (Berki et al., 2004) and software quality depends on the opinions and attitudes of the stakeholders, who are concerned

Table 1. Values in traditional SPI and in agile paradigms

SPI	Agile
Processes and tools	Individuals and interactions
Comprehensive documentation	Workable software
Contract negotiation	Customer collaboration
Change through following a plan	Change through fast response

Table 2. Quality factors in the basic principles and values of SPI and agile paradigms

Quality factors	SPI	Agile
Philosophy	Empowerment	Innovative, participative, empowerment
Lifestyle	Work-orientated	Life-orientated
Approach	Plan driven and prescriptive processes Process driven--rigid--bureaucratic	Flexible Evolutionary, adaptive, iterative, incremental
Driving forces	Management commitment and leadership (Deming, 1986)	Technically competent and motivated developers
Customer involvement	Early and late stages in life cycle	Throughout life cycle
Customer participation	Encouraged—Customer Focus	Imperative user participation
Communication	Formal	Informal
Teams	Inter-group coordination (Humphrey, 1995; Deming, 1986)	Self-organising teams
Responsiveness	Bureaucratic delays	Quick responses
Knowledge creation	Tacit, Formal, Explicit	Tacit, Informal, Explicit
Knowledge sharing	Desirable, Formal	Imperative, Informal
Documentation	Maximum	Minimum
Changing requirements	Processes have to be followed	Adaptability to changes throughout the development process (Berki, 2001; 2004)
Testing	Late in life cycle	Test first (Holcombe, 2005)
Error detection	Inspection (Gilb & Graham)	Pair programming (XP) (Beck, 2001; 2003)
Progress reviews	Formal peer reviews (CMMI)	Continuous peer reviews
Requirements elicitation	Planned and infrequent	Daily stand-up meeting
Tool support	Tools supporting different phases of the life cycle—fragmented	Automated testing tools—Integrated CASE (I-CASE)
Delivery of product	Planned	Frequent—loose plan
QA function	Formalised—Separated	Informal—Embedded

with different quality attributes (Siakas & Georgiadou, 2005; Siakas et al., 1997). Quality is in the eye of the stakeholder!

Evidence for the emphasis on process quality is also that ISO certification does not certify product quality but takes into consideration that a stated and documented process is followed. Similarly, well-known and recognised assessment models like CMM / CMMI (CMM, 2005; Paulk, 1993, 1995; Paulk, Curtis, & Chrissis, 1993), BOOTSTRAP (Haase, 1992; Haase & Messnarz, 1994; Kuvaja, Similä, Kranik, Bicego, & Saukkonen, 1994), and SPICE / ISO15504 (Dorling, 1993) concentrate on the assessment of the process quality and not on the quality of the final product.

Critically reflecting and summarising on the values and principles of traditional software process improvement (SPI) and agile paradigms, Tables 1 and 2 provide a comparative review on the trends and shifts that the two approaches are based on. Firstly, Table 1 outlines the values of SPI and agile approaches. Table 2 continues to expose the quality factors and the special features/principles through which quality is encapsulated for the software development process, and/or is built in for the software product, being that a workable prototype—a frequent release or the final product itself.

AGILE METHOD(OLOGY): A NEW TREND WITH LITTLE QUALITY ASSURANCE OR INHERENT QUALITY FACTORS?

Apart from the quality properties of implementability, changeability, and testability, agile methods have re-introduced a stakeholder—or user—centred approach relating to requirements conformance, concentrating on the software/information products' frequently released versions in order to ensure competitive advantage, end-user acceptance, and general stakeholder satisfaction. Factors, though, that affect the quality of software (and their interconnected nature) need to be identified and controlled (Georgiadou et al., 2003) to ensure predictable and measurable totality of features and quality characteristics of the particular software.

There is significant evidence that culture influences on how people perceive quality and quality assurance (Berki, 2006; Ross et al., 1995; Siakas et al., 1997), and that there are differences amongst IS professionals from different nations (Couger & Adelsberger, 1988; Couger, Borovitz, & Zviran, 1989; Couger, Halttunen, & Lyytinen, 1991; Holmes, 1995; Ives & Järvenpää, 1991; Keen, 1992; Kumar & Bjorn-Andersen, 1990; Palvia, Palvia, & Zigli, 1992, Siakas, 2002). Revisiting this chapter's main research question: Does the much praised applicability of agile methods fit different organisational and national cultures and human needs in a world of turbulent quality perspectives under the globalisation of software development? The following section considers this question in the context of measurement and other factors.

WHAT TO MEASURE AND HOW TO IMPROVE QUALITY IN AGILE SOFTWARE DEVELOPMENT?

It has long been recognised that measures provide insights into the strengths and weaknesses of both process and product. *To measure is to know. If you cannot measure it, you cannot improve it* (William Thomson, later Lord Kelvin (1824-1907); DeMarco, 1982). Hence, the "why to measure" is understood and accepted, but the decision on "what to measure," "how to measure," "when to measure," and, in particular, "who measures" and "for whom" becomes crucial if benefits are to be gained by particular groups (Berki, 2001). In turn, we need to understand which stakeholders and why they will benefit from measurements and improvements, and in the long run, who is the most suitable to measure and bring about reforms, changes, and improvements in the software products and processes. The following two sections take a closer look at the role of measurement and estimation in traditional software development while the two sections after them intend to introduce the concepts and attributes of measurement and estimation, regarding process and product quality improvement, in agile software development.

The Role of Software Measurement in Software Quality Assurance

Hennel (1988) and Fenton (1991) believe that *internal attributes* are the key to improving software quality and can be measured in terms of the code. Many traditional software engineering methods provide rules, tools, and heuristics for producing software products. They show how to provide structure in both the development process and the products themselves such as documents and code. These products have properties (internal attributes) such as modularity, re-use, coupling, cohesiveness, redundancy, D-structuredness, and hierarchy. They aim to assure reliability, maintainability, and usability for users, and also assure high productivity and cost-effectiveness for managers/sponsors.

External attributes (Fenton, 1991; Hennell, 1991; Georgiadou, 1994) such as understandability and maintainability are behavioural, and their associated metrics are both qualitative and quantitative. They are always obtained indirectly through the use of surrogate measures. In general, product measurement affords us with direct metrics (in terms of counts such as NLOC and ratios such as density of calls), whereas process measurement invariably takes the indirect route of surrogacy and use of indicators and indices (Georgiadou, 1994, 2001; Kitchenham, 1996).

In the light of the previous and examining the relevant direct, indirect and surrogate measures that are kept or/and could be kept in agile paradigm methodologies, Table 3 characterises the nature of these measurements.

The Role of Cost Estimation Models in Traditional ISD

In traditional software development, cost estimation models such as the original COCOMO (COnstructive COst MOdel) (Boehm, 1981) and subsequent versions (COCOMO Web site, 2006) aimed at estimating the cost, effort, and schedule when planning new development activities, according to software development practices that were commonly used in the 1970s through the

Table 3. Measurement in agile software development

Agile	Measures: Direct/Indirect/Surrogate
User satisfaction	Surrogate
Changing requirements	Direct
Frequent delivery of working software	Direct
User participation	Indirect
Developer motivation and trust	Indirect/Surrogate
Efficiency through face-to-face conversation	Indirect
Working software—the primary measure of progress	Indirect
Sustainable development	Surrogate
Continuous attention to technical excellence and good design enhances agility	Indirect
Simplicity (delivery of essentials functionalities)	Surrogate
Self-organising teams	Surrogate
Regular reflection and tuning of teams behaviour	Indirect

1980s. COCOMO and its variations link internal and external attributes (see earlier section) providing rough estimates of software costs. Accuracy is necessarily limited because of the difficulty to account for differences in hardware constraints, personnel quality and experience, use of different tools and techniques, and other project attributes that may have a significant influence on cost. Effort-predicting metrics must inevitably be available fairly early on in the development life cycle if management teams are expected to commit expenditure and plan the timing of delivery. In addition, Beizer (1990) observed that it is important to be aware of the purpose of a *metric;* moreover, confusing an *effort-predicting metric* with a *defect-predicting metric* can nullify the metric's usefulness.

Product or Process Quality Measurements in Agile Software Development?

Could flexible agile quality metrics exist for the agile software development process and its deliverables? The agility here refers to the ability to continuously modify both cost and effort using parameters such as function points and lines of code. For instance, a recent variation of COCOMO and associated Web-based tools is the Agile COCOMO II, which facilitates cost estimation and enables adjustments to "estimates by analogy through identifying the factors that will be changing and by how much" (Agile COCOMO Web site, 2006).

Gilb (2006) suggests that the agile methods paradigm would benefit, if it included "stakeholder metrics" for focusing on critical requirements, measuring progress, and enabling response to feedback. In suggesting so, Gilb (2006) asserts the need for agile development to focus on the most critical stakeholder quantified and measurable requirements and the need to keep track of changes and plans for better progress in shorter project cycles. However, rigorous and flexible agile metrication procedures could be difficult to establish for the requirements of agile software development. Evidentially there is no concrete, established scientific procedure to do so and, furthermore, neither vast industrial nor academic substantial proof for keeping quantifiable measures in agile software development projects.

Some agile metrics, for instance, have been reported and used by a research team in the Israeli army. The reported results support the stated goal, that is to "enable an accurate and professional decision-making process for both short- and long-term purposes" (Dubinsky, Talby, Hassan, & Keren, 2006).

On the other hand, agile metrics are not themselves very different from traditional metrics (see also Table 3). The difference in "agility" probably refers to the method and speed of collection. Since, though, agile methods advocate lean documentation and frequent delivery of product, agility in estimation and continuous adjustment is necessary so that projects are delivered on time and within budget. In order for this iterative process to succeed, we must assume the practice of a *mature* process within the organisation.

At capability maturity model (CMM) level 5—the optimising level—quantitative feedback data drawn from the process allows continuous process improvement. At this maturity level (Humphrey, 1995; Zahran, 1998), data gathering must be automated and the emphasis should be shifted from product maintenance to process analysis and improvement. Managing the agile development process, and any process, requires the collection of suitable metrics, which will provide holistic insights into the strengths and weaknesses of the process, assuring general stakeholder satisfaction.

Figure 3. The evolution of the life cycle model maturity (Georgiadou, 2003b)

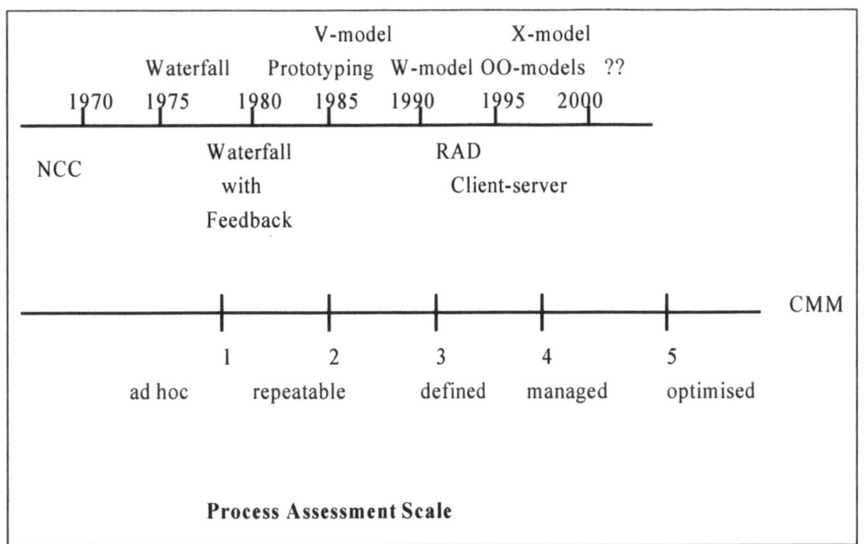

TESTING, IMPROVEMENT, AND RE-ENGINEERING FACTORS IN TRADITIONAL AND IN AGILE LIFE CYCLE PROCESSES

Efforts to model the process gave us numerous life cycle models, ISD methodologies, quality models, and process maturity models (Berki et al., 2002). A study of different life cycle models was carried out (Georgiadou, 2003b; Georgiadou et al., 1995) with particular emphasis on the position of testing in the whole of the life cycle. The study so far recognised that more mature life cycles, and hence more mature software engineering processes, moved testing and other quality assurance techniques to the earliest stages of the development life cycle.

A juxtaposition of a historical introduction of the models to the CMM scale (Figure 3) demonstrates that between 1970 and 2000, as we moved from the waterfall model (which appeared as early as in 1956) to incremental models, the maturity of the underlying process was increasing. In the case of agile development methods, the testing activities are frequent and central for stakeholder satisfaction. There are even approaches with stakeholder requirements testing before development, hence, the maturity of the software development process continuously raises with the awareness and advanced practices of testing!

Nevertheless, cyclic development models, incremental prototyping (Pressman, 2000; Shepperd, 1995), and rapid application development (RAD) (Bell & Wood-Harper, 1992) offered mechanisms for monitoring, timing, and quality of deliverables. Prototyping maximizes the probability of achieving completeness because of the principle of delivering workable versions upon user acceptance. The need for continuous improvement was widely recognized, and is encapsulated by the concept of Kaizen ("Kai" meaning school and "Zen" meaning wisdom), which is a Japanese word meaning "many small improvements" (Imai, 1986).

According to Beck (2000, 2001), traditional lifecycle models are inadequate and should be replaced by incremental design and rapid prototyping, using one model from concept to code through analysis, design, and implementation. In other words, design starts while still analysing and coding starts soon after starting the detailed design. Portions of the design are tested along with implementation. Extreme programming (XP), for instance, is one of the agile or lightweight methodologies to denote a breakaway from too many rules and practices.

The XP approach is considered successful since it stresses customer satisfaction upon design, testing, implementation, subsequent and on-going customer participation, and feedback. The particular methodology was designed to deliver the software a customer needs when it is needed. XP (and most other agile methods) empower developers to confidently respond to and test changing customer requirements, even late in the life cycle (Beck, 2002; Holcombe, 2005). Acceptance of changes also empowers the customers and end-users, but do additional requirements violate the requirements specification and do they invalidate carefully tested and documented requirements within previous modelling efforts? Unless, of course, the continuous re-structuring and re-engineering of requirements specification empowers reverse engineering and re-factoring. Refactoring, an agile concept and different from rearranging code activity, whereupon the strength of testing lies therein, perhaps indicates that frequent requirements change doesn't preclude changing functionality. The following sections take a closer look to re-engineering, restructuring, reverse engineering and refactoring with reference to the agile development paradigm.

Software Design Quality Factors

We are interested in both internal and external quality attributes because the reliability of a program is dependent not just on the program itself, but on the compiler, machine, and user. Also, productivity is dependent on people and project management. It is often necessary to use surrogate measures for complex external attributes (Fenton & Pfleeger, 1997; Kitchenham, 1996).

Legacy code accounts for more than 80% of the total effort in the whole of the community (Polymenakou, 1995). The extent to which the industry is still using 30 or even 40-year-old systems became all too apparent with the Millennium bug, which was an unintended consequence of much earlier design decisions and testing limitations. Maintenance of existing code requires understanding of the requirements and the behaviour of the system as well as of new technologies, such as new database management systems, new programming languages, new methods, and tools that cater for change and re-use. Legacy code may need drastic restructuring, and often reverse engineering due to inadequate or absent documentation to facilitate re-engineering.

The Four Rs: *Re-Engineering, Re-Structuring, Re-Use, Re-Factoring*

When considering software development we generally think of a natural process starting from the concept, going through the specification, design, implementation, and delivery of product (going live) which in turn reaches obsolescence. This process (modelled by different life cycles) is known as *forward engineering*. In contrast, *reverse engineering* starts from a product and in reverse order generates specifications (Pressman, 2000). Legacy code without accompanying documentation (or with documentation, which has long ceased to keep in step with modifications) is often a serious problem. The dilemma is whether to reverse-engineer or to embark onto development from scratch. Cost and time constraints play a significant part in the resolution of this dilemma.

A more frequent activity is the re-engineering or restructuring of code for the purpose of improving its performance and/or as a result of changing requirements and hence, sometimes, changing functionality. Here, we need to at least ensure the preservation of the existing functionality (unless parts of it are not required) and to also enhance problematic areas and add new functionality as required. Re-engineering of existing code needs to be planned and managed according to specified rules of acceptance thresholds and tolerance. Again, the costs of reengineering may be just as high if not higher than the costs for forward engineering (developing from scratch).

Reuse has long been advocated as the key to productivity improvement. Fenton (1991) specifies *private reuse* as the extent to which modules within a product are reused within the same product whilst *public reuse* is the proportion of a product, which was constructed (Fenton, 1991). The problem with reuse is whether a system is reusable. Often it is necessary to restructure existing code before reuse can be made possible. Funding a reuse program can be expensive, and it is difficult to quantify the cost reductions expected from reuse (Sommerville, 2001). It is desirable that components (modules) should be highly cohesive and loosely coupled with other components. The authorised re-user of the components must have confidence in that the components will behave as specified and will be reliable. The components must have associated documentation to help the re-user understand them and adapt them to a new application. In recent years, much promise came from the OO paradigm and component-based development (CBD).

Agile methodologies, and XP in particular, introduced the concept of re-factoring as a process of "changing a software system in such a way that it does not alter the external behaviour of the code yet improves its internal structure" (Fowler, 1999). Fowler advocates that re-factoring is typically carried out in small steps. After each step, you are left with a working system that is functionally unchanged. Practitioners typically interleave bug fixes and feature additions between these steps. In so doing, re-factoring does not preclude changing functionality; it supports that it is a different, from re-arranging code, activity.

The key insight is that it is easier to rearrange the code correctly if you do not simultaneously try to change its functionality. The secondary insight is that it is easier to change functionality when you have clean (re-factored) code. The expectation is that re-factored code will be more maintainable, which in turn, will result in efficiency and cost savings. The cyclic nature of re-factoring also improves understanding in the development team, and ensures the reuse of tried and tested code. Re-factoring, therefore, improves the overall quality of software.

Notwithstanding, agile methods have, to-date, failed to propose measures for process or product improvement. Additionally, one can envisage cost over-runs, which can neither be predicted nor controlled. The re-engineering metrics developed by Georgiadou (1994, 1995, 1999) provide indicators as to whether legacy code can be restructured or developed from scratch. Re-factoring is in effect a re-engineering activity. If we hope to control the costs and time of delivery, benefits could be gained from adopting metrics such as the re-engineering factor *rho* (ρ) (Georgiadou, 1995) and the associated productivity gains in agile development.

NON-MEASURABLE AND UNCONTROLLABLE QUALITY FACTORS IN AGILE SOFTWARE DEVELOPMENT

In knowledge intensive organisations, strategic decisions should be based on the development and improvement of their own resources (Conner & Prahalad, 1996). This also means development of a flexible organisational culture, which can quickly respond to various business situations (Rantapuska, Siakas, Sadler, & Mohamed, 1999). One of the basic features in agile methodologies is to embrace changing requirements with a quick response. No detailed plans are made. Instead, changing requirements is considered a necessity to sustain and improve the customers' competitive advantage.

The previous meta-requirements of agile software development can be critical and uncontrollable quality factors while developing human-centred software for technological products and organisational, societal, personal, and interpersonal information systems (see e.g., Koskinen, Liimatainen, Berki, & Jäkälä, 2005). When designing huma-centered systems requirements flexibility as well as the debate between changeability vs. stability of the object domain of software are central questions for quality considering different stakeholders intersts and values. The following section briefly refers to the research findings on recent quality measurements made for software development process in European countries and summarises the conclusions from previous published work by the authors of this chapter. The three subsequent sections scrutinise the concepts of knowledge sharing, cultural factors, and stakeholder participation as soft issues and unpredictable software quality factors within the agile paradigm.

Current Practices and Measurements Kept for Software Quality: A European Perspective

In a comparative study carried out in the form of triangulation (quantitative and qualitative investigation) in four counties, namely Denmark, Finland, Greece, and the UK, cultural differences in software development and their influence on software quality were investigated (Georgiadou et al., 2003; Siakas, 2002; Siakas et al., 2003; Siakas & Balstrup, 2000; Siakas & Georgiadou, 2002). The questionnaire was sent to organisations developing software for own use or for sale, and normally kept formal measures of software quality. In total, 307 questionnaires were completed. In addition, field-studies were undertaken in several organisations. In total, 87 interviews were conducted in Finland, Denmark, and Greece with software developers at different levels and with different positions in the organisations. Following the initial verification phase, observations were carried out in a Danish organisation for a period of two months (Siakas et al., 2000). The objective of using observations was to investigate in more depth the research problem of organisational and national culture diversity and their influence on software quality management practices and to verify the findings. Amongst other findings, we proved that there are statistically significant differences in the responses on the degree to which *formal* quality measures are kept depending on the country of origin. Figure 4 depicts some of the research findings.

From Figure 4 we observe that amongst the software development organisations taking part

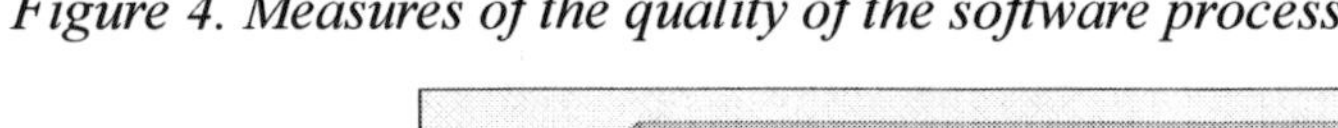
Figure 4. Measures of the quality of the software process

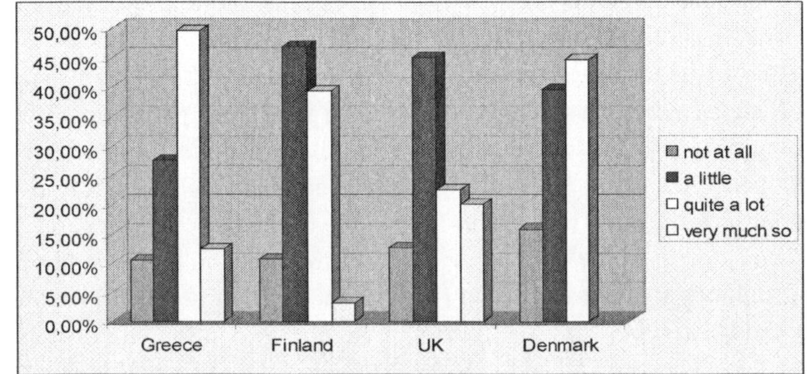

in the study, Greece is the country that keeps measures of the quality of the software development process to the highest degree. The sum of the values for "quite a lot" or "very much so" is 61.9% for Greece, 44.7% for Denmark, 42.6% for Finland, and 42.5% for the UK. The significance of the Chi-square is 0.002, which indicates that the null-hypothesis, that the responses are similar for all countries, can be rejected. This means that we have statistically proved that there are significant differences in responses depending on country of origin.

The research study itself was the starting point to examine and classify in a typology the different organisational and national cultural factors and their influence in software quality management strategies. In what follows we draw from these conclusions and attempt their re-framing and examination within the agile paradigm principles and strategies for stakeholders' involvement.

Knowledge Creation and Transfer in Agile Development

Knowledge is an active part of human working practice. Malhotra (1997) lists four concerns regarding knowledge creation in organisations, namely:

- Dynamic and continuously evolving nature of knowledge.
- Tacit and explicit dimensions in knowledge creation.
- Subjective, interpretative, and meaning making base of knowledge creation.
- Constructive nature of knowledge creation.

Knowledge evolves continuously as the individual adapts to amplifications and innovations from their peer workmates. Working knowledge is at a great proportion tacit, which means that people rely on their earlier experiences, perceptions, and internalised knowledge instead of expressing knowledge by explicit procedures. Successive knowledge creation also requires voluntary actions including openness, scrutiny, and reluctance to different views and interpretations.

New knowledge always begins with the individual. The individual's self-interest determines in which informal knowledge creation processes to participate (Chen & Edgington, 2005). Individuals also tend to hoard knowledge for various reasons (Bock, Zmud, Kim, & Lee, 2005); one reason being the cultural value system of the team, organisation, or country (Siakas & Mitalas, 2004). Within a single culture certain beliefs, values, procedural norms, attitudes, and behaviours are either favoured or suppressed (Siakas et al., 2003). Making personal knowledge available to others is a central activity in knowledge creation. Explicit knowledge is formal and systematic, while tacit knowledge is highly personal (Nonaka, 1998). The constructive learning process in work places is revealed through bottom-up knowledge creation spread from individual to individual in the socialisation process (Nonaka & Takeuchi, 1995).

Because of extensive customer involvement in the agile development process, and the direct feedback through frequent face-to-face communications between customers, representatives, and developers, there is an efficient transfer of individual knowledge. Pair-programming, pair-rotation across programming pairs and feature teams, and peer-reviews could also facilitate tacit knowledge creation and transfer (Boehm & Turner, 2003a, 2003b; Luong & Chau, 2002). Organisations and teams expect their members to keep up-to-date by continuously obtaining internal and external information relating to their profession. Team members reflect in action, accumulate, and refine the knowledge required for completing their tasks (context) through a process of learning-by-doing (Garud & Kumaraswamy, 2005). Initial tacit knowledge created in the working practice is only the first step in the organisational knowledge creation process according to Nonaka et al. (1995). In addition, knowledge is sometimes introduced into organisational use by making it first explicit and then transferring back to the members of the organisation. In agile methodologies, this part seems to be disregarded. Individuals may maximise personal benefit, but to transform tacit knowledge into organisational knowledge (if possible) is not an easy task, because tacit knowledge is human capital that 'walks out the door at the end of the day" (Spiegler, 2005).

In order to utilise the individual knowledge and transfer it into organisational knowledge, the organisation needs to strategically align organisational learning investment with organisational value outcome taking into account both current and future organisational tasks. The organisation has more control over a formal or structured knowledge creation process (Chen et al., 2005). However, deliberate knowledge management (KM) involves more than sponsorship of initiatives at and across different organisational levels (Garud et al., 2005). It also involves an active process of causing intended consequences. The organisational culture, including the organisational structure, processes, and procedures, is important for avoiding organisational knowledge dilution, creating, transferring, and sustaining organisational knowledge. Recently, there were indications for an emergent need for integrating KM and agile processes (Holz, Melnik, & Schaaf, 2003).

In agile development, documentation is kept to a minimum (Agile manifesto, 2005). Documentation is seen as a non-productive and non-creative task resulting in static documents that seldom reflect reality and are hardly ever used after they are created. The question seems to be how much return on investment (ROI) and added business value are provided by documentation. Agile development focuses on executable documentation, also called agile testing (Pettichord, 2002), namely self-documented code, which is code including comments, self-describing variable names, and functions, as well as test cases, needed to accomplish the project. Some high-level conceptual documentation such as user stories (short descriptions of features in requirements), high-level architectural design, and/or documentation of high-risk components are usually also created to define a contract model or to support communication with an external part (Amber, 2005). This type of documentation accomplishes a high-level road map, which can be useful for projects newcomers and post-mortem analysis.

The close collaboration of team-members, including stand-up meetings, peer-reviews, pair-rotation, and pair-programming in agile development ensure knowledge of each other's work and therefore improve domain knowledge and quality of the deliverables. Automated documentation tools, such as design tools, may be used for the final documentation, while during the development frequent iterations would cause version problems (Manninen et al., 2004). Research has shown that there is a visible conflict related to the amount of documentation that should be kept (Karlström & Runeson, 2005); a balance between how much work to put into documentation and the usefulness of documentation has to be found.

Customer Involvement vs. End-User(s) Representative or Consensus Participation

The distinction between predictive (plan-driven) and adaptive (agile) recognises the fact that achieving plan milestones does not necessarily equate to customer success (Mellor, 2005). In order to ensure conformance to requirements, user satisfaction, and competitive advantage agile development involves the user in the entire process. However, customer identification can be difficult, and may require the identification of suitable internal customer representative(s) providing a single point of contact both for the team and senior management on a daily basis. The high iteration frequency in agile development also provides opportunities for product feedback on conformance to requirements

(Karlström et al., 2005). The emphasis on user viewpoints relating to the characteristics of the final product (user-centred approach) in combination with a daily feedback mechanism increases the rate of feedback on performed work and the speed of discovering erroneous functionality at an early development stage (Siakas et al., 2005; Karlström et al., 2005).

Developer motivation, commitment, and satisfaction are key elements for success (Abrahamsson, 2002; Siakas et al., 2003) due to the fact that they recognise the actual need of a quality outcome and to the importance of their role in the creation process (feelings of personal ownership). The high iteration frequency also has consequences for contracts variables, such as scope, price and time and thus the contracts need to be flexible (Geschi, Sillitti, & Succi, 2005). This, in turn, may be a drawback for the customer's cost-analysis plans.

Organisational and National Culture Issues: Is Agility Acceptable and Suitable for all Cultures?

The basis of agile development lies in small teams working in co-located development environments developing non-safety critical software (Abrahamsson, 2005). Agile development relies on self-directed teams consisting of highly skilled, motivated, and innovative software engineers, who are collaborative in team work and self-organised, active learners. These characteristics impose a competitive environment with potential cultural, political, and social implications. Integrating agile approaches and philosophies into traditional environments with existing "legacy" staff and processes is difficult (Boehm et al., 2003a, 2003b). The focus on agility and simplicity, people-orientation, and final product delivery in agile development indicates many degrees of freedom and individual skills building. This is opposed to process orientation and maturity approaches through frameworks of policies and procedures, organisational rules, and regulations that empower developers and technical staff by giving a back-to-basics sense to their work (Boehm et al., 2003a, 2003b; DeMarco & Boehm, 2002; Karlström et al., 2005). Competent agile teams that possess the necessary mix of technical skills, people expertise, and agility are built on cohesive team environments and are committed to the common goals. However, generating group identity may prove difficult (Boehm et al., 2003a, 2003b). The main issues though seem to be changes in values and attitudes in organisational culture and function and synchronisation of teams based on communication, leadership, and trust (Jäkälä & Berki, 2004). The larger the organisation and the more traditional, the more difficult is a cultural change in the organisation (Siakas, 2002).

The agile approach to software development has the characteristics of a group of people that differentiate themselves from others through a whole set of shared values, visions, principles, ideals, practices, etc, that emerge in the interaction between members of a group. The extreme programming (XP) pioneers for example, draw attention to four XP values, namely, communication, simplicity, feedback, and courage, the underlying basis for the 12 principles which are translated into practices (Robinson & Sharp, 2003). These practices are the artefacts of the XP culture.

Having a consistent culture is important for consensus and agreement, as well as for avoidance of friction due to cultural clashes within the team and the whole organisation. The social characteristics of the team members are important. Employment of technically highly competent and competitive software professionals generates the basis for the creation of a strong team/organisational culture. The agile culture requires active involvement of all team members and seems to be most suitable in Democratic-type of organisations, which have horizontal hierarchy emphasising flexibility and spontaneity (Siakas et al., 2000). This type of organisation generates initiative and responsibility approaches; the leadership style is that of coordination and organisation. The organisation has flexible rules and problems are solved by negotiations. Employees are encouraged to make contribution to the decision-making process and to the development of the organisation in general. Democratic organisations can be said to be people-oriented. Examples of countries belonging to the Democratic type are some Scandinavian countries, Anglo-Saxon countries, and Jamaica (Siakas et al., 2000).

The globalisation of software development, involving virtual teams and offshore outsourcing, increases complexity. The management of large, global organisations, dependent on people with different underlying norms, values, and beliefs, experiences difficulties when applying traditional management approaches (Siakas et al., 2005). Despite significant improvements in ICT, the complexity will increase even more if agile software development is introduced in such global settings. By nature, agile development facilitates pair-programming, pair-reviews, and pair-rotation, and thus, could not be facilitated among virtual teams. Additional national differences in working values, attitudes, and time-zone differences will be difficult to overcome.

Having to consider so many differences and desires for empowerment of developers and end-users, suitability and applicability of an agile method in a particular organisational, cultural, or national setting still remains questionable. Moreover, method customisation and adaptability issues are not completely resolved for maximum designer and end-user involvement. XP (and most other agile methods) empower developers to confidently respond to and test changing customer requirements, even late in the life cycle (Beck, 2002; Holcombe, 2005). Acceptance of changes also empowers the customers but do additional customer requirements violate the requirements specification and do they invalidate previous modelling efforts? Unless, of course, the requirements specification is empowered by a method which is flexible, customisable to designers and to end-users needs. This, of course, moves the problem to a more abstract level, that of *method metamodelling* and *method engineering*. The next section explores these issues further in the context of this chapter and of this book.

THE NEED FOR FORMAL METAMODELLING AND AGILE METHOD ENGINEERING PRINCIPLES WITH SUITABLE TOOL SUPPORT

Assuming that the benefits of using agile methods are remarkable and unquestionable, how could practitioners and software project managers choose among agile and non-agile methods? How could someone evaluate the "agility" or "agileness" of an agile method as this is defined in the agile manifesto (2005) (Cockburn, 2004), and as is defined by many other supporters? Moreover, how could systems analysts and designers construct and/or adopt an agile method if they are not fully satisfied with the ones currently available? Otherwise, how could the quality properties of requirements changeability (and therefore specification modifiability), implementability (and therefore specification computability), and frequent deliverable artefacts' testability be established and assured? The answer(s) probably lie in utilising cost-effective technology and existing MetaCASE tools (Berki, 2004). Implementability, changeability, and testing, as hinted in the agile manifesto and as praised by the agile methodology supporters, are considered as the significant difference quality properties offered in agile software development teams. Moreover, and turning to soft issues of collaboration and communication between end-users and IS designers, how could the properties of communicability and synergy among stakeholders be ensured for the agile method to be used?

Because method, application, software, and process engineers want to create their own knowledge-oriented methods, document their own tool environment, and design their software development processes, they frequently resort to metamodels and method engineering (ME) principles (Berki, 2001). In so doing, assurance is granted for customised and flexible methods and for maximum designer involvement and user participation (Berki, 2004, 2006; Georgiadou et al., 1998). Bridging the gaps that traditional non-agile design methods created, the utilisation of MetaCASE and computer-assisted method engineering (CAME) tools (Berki, 2001; Tolvanen, 1998) for the evaluation and/or construction of an agile process model (method) could be the answer to the quality assurance required for agile methods.

Agileness and its Meaning in terms of Method Engineering

Representing the rules of metamodelling and method engineering in the context of MetaCASE (CASE-Shells) and CAME tools, a proposal to model agile and formal process models in CAME environments requires to utilise meta-design knowledge patterns that facilitate collaboration and interaction. In order to construct an agile method, many quality properties such as the ones previously mentioned are required.

Method flexibility and adaptability to accommodate changes is of foremost importance. When requirements, specifications, and programming artefacts undergo changes after frequent end-users feedback, methods' models and techniques need to be extended and integrated in order to facilitate the accommodation of new stakeholders' needs. Expressing and understanding method techniques and models in terms of their dynamics requires the utilisation of formal meta-concepts to define new representations of a method syntactically and semantically. Furthermore, a generic and agile method model should possess computational characteristics for facilitating frequent future implementations and their redesign in a testable way (Berki, 2001).

The Need for *Agile Method Engineering (AME)* with CAME tools

The expansion and use of agile methods and the associated technology gave rise to more demanding modelling processes among the ISD stakeholders. On the one hand, the need for preserving the consistency, correctness, and completeness of the associated artefacts of every developmental stage indicated the need for advanced software and groupware tool use and shared and agreed work and method models for communication and collaboration. On the other hand, more abstraction, analytical, and communication skills are constantly required to collaborate with a variety of stakeholders. These integrated requirements assisted in realising the need for more cooperation and interaction in the IS life cycle and, at the same time, the need to capturing the computational characteristics and the dynamic knowledge of a method. The diverse cognitive needs of human behaviour and interaction (Huotari & Kaipala, 1999), in particular, gave rise to expressive software systems modelling through shared process models that represent agreed views. People change and their opinions and requirements constantly change, too. Systems engineers that use agile methods need evolutionary and expandable models to capture modelling concepts and communicate them with others (Berki, 2004).

Existing MetaCASE technology offers to humans possibilities to formalise the systems development processes and lead to standardisation and integration of agile methods and tools with quality assurance procedures. The great challenges though of MetaCASE technology will be to provide guidance on implementation strategies and facilitate the requirement for testing of agile modelling processes and products. Some of the MetaCASE tools that can be used to construct agile methods in a dynamic and computational way are CoCoA (complex covering aggregation), MetaView, NIAM (Nijssen's information analysis method), and Object-Z. Not all, though, are fully automated for complete use as MetaCASE tools. Moreover, they do not offer method testability (Berki, 2004, 2006). The exception is MetaEdit+, a multi-method, and a multi-tool CASE and CAME platform, which offers the modelling power to construct an agile method with its method workbench to varying degrees of support (Kelly et al., 1996; Tolvanen, 1998;).

A Formal Method Workbench to Built-In Quality When Constructing Agile Methods

MetaEdit+ establishes a versatile and powerful multi-tool environment that could enable flexible creation, maintenance, manipulation, retrieval, and representation of design information among multiple developers. MetaEdit+ is a multi-user tool constructed to involve as many possible developers and multi-users in a flexible creation of, a suitable to *their* needs, method. The tool has been used worldwide in academia and in industry but it has mainly been utilised in systems development in Scandinavian countries, that is the democratic type of countries (see previous section entitled: "Organisational and National Culture Issues: Is Agility Acceptable and Suitable for all Cultures?").

Being of such nature and created to address the needs for participation and maximum user involvement, MetaEdit+ could point to and appraise the suitability and applicability of an agile method with respect to stakeholders' needs. As a computer aided method engineering (CAME) environment with a method workbench, it offers an easy-to-use, yet powerful environment for agile method specification, integration, change management, and re-use (Kelly, et al., 1996; Rossi, 1998; Tolvanen, 1998). The tool could also facilitate method implementations in a computable manner and could provide capture of frequently changing requirements and testing (Berki, 2001, 2004), both requirements of foremost importance in agile systems development. For instance, at MetaEdit+ an agile method (and any method) could be represented as a *Graph, Table, or Matrix*, having the following semantic and syntactic meta-design constructs:

- **Graphs:** Sets of graphical objects and their connections.
- **Objects:** Identifiable design entities in every technique/method.
- **Properties:** Attributes of graphs, objects, relationships, and roles.
- **Relationships:** Associations between objects.
- **Roles:** Define the ways in which objects participate in specific relationships.

The method workbench is a significant part of MetaEdit+ tool. The basic architectural structure that is used to create the products of all levels (i.e., methods and their instances) is GOPRR. As outlined earlier, GOPRR recognises in a method's generic structure (and therefore in its instances) the semantic concepts of objects and relationships, which both possess properties and roles. When creating new method specifications in MetaEdit+, the metamodeller should firstly concentrate on the constructs of a method (Kelly, et al., 1996; Rossi, 1998; Tolvanen, 1998). In doing so, he or she must use the GOPRR metamodel to guide the whole metamodelling process as well as the assisting drawing, hypertext, etc. tools that are offered with MetaEdit+ and are depicted in the MetaEdit+'s architecture, which is presented in Appendix B.

Naturally, this way of metamodelling offers a degree of integration but limited expressability for data and process dynamics. Methods are process models that transform data continuously but very few dynamic methods exist in MetaCASE tool support (Berki, 2004; Tolvanen, 1998). And, finally, formalised generic guidelines of a "process for modelling an agile method" together with formal testing procedures and suitable method engineering metrics should be incorporated in the meta-design process of the next generation of MetaCASE and CAME tools. Furthermore, hypertext facilities are under development in order to express explicit knowledge (knowledge about method's constructs) more accurately but also facilitate the expression of tacit knowledge, that is the feedback mechanisms and opinions among stakeholders at many different levels.

The Need for an Agile, Formal, Generic Metamodel

It is argued that computability and implementation issues in method metamodelling can be captured with more abstract and classic computational models (Berki, 2001; Holcombe & Ipate, 1998). With this choice, the method's computational quality properties can be documented and communicated as implementation details for the system designers, when the purpose is to analyse the given stakeholders needs and design and implement them into a software system. Such a facility can be offered by the CDM-FILTERS model (Berki, 2001), which provides an integrated specification platform for agile software development with the agile paradigm qualities and values (see again Tables 1 and 2).

CDM-FILTERS stands for a Computational and Dynamic Metamodel as a Flexible and Integrating Language for Testing, Expressing, and Reengineering Systems. It was constructed as a result of a large survey in metamodelling technology and existing MetaCASE and CAME tools and as a promise to overcome severe limitations that were observed in software development with traditional information systems development methodologies in the last 40 years (see Figure 1). As a metamodel, CDM-FILTERS is based on machines (general finite state machines) that provide the inherent quality properties of changeability as dynamic models and specification computability by deriving implementable designs. Moreover, machines are testable computational models, known for the *finite state machine standard procedure for testing* in early phases of specification and design. As a framework, CDM-FILTERS recognises the evolutionary and frequently changing nature of systems and their stakeholders and facilitates standard feedback and communication mechanisms among them. (Berki, 2001, 2004).

Thus, with CDM-FILTERS as an agile method engineering instrument, it will be possible to evaluate and integrate existing methods and build and test methods by revealing and testing at the same time the implementation details of the prototype code that needs to be frequently released. This *conceptual computational* modelling, which is hereby suggested for agile software development processes, is based on *dynamic metarepresentations* of methods, and is achieved by capturing the method's *pragmatics*, semantics, and syntax as *machines* specifications, which are general, dynamic, and testable computational models (Berki, 2001, 2004). This *systemic* (holistic) metamodelling and process-based approach can offer a smooth transition between the systems development phases by improving the communication channels for stakeholders in requirements engineering and re-engineering, and by mapping the exact testable artefacts of one stage to those of the next (Berki, 2001), which is a major challenge in traditional software systems engineering and in the agile software development processes.

DISCUSSION, CONCLUSIVE REMARKS, AND FUTURE RESEARCH CONSIDERATIONS

In identifying the reasons for the popularity of agile methods and state-of-the-art in software engineering, we examined the quality of agile software development processes and analysed socio-technical and organisational factors related to quality information systems and stakeholders' involvement and satisfaction. Contemporary agile methods, compared to traditional software development methods, still need to demonstrate further practicality, applicability, their "generic" suitability, and finally, their potential for quality assurance. Some research projects from industry and academia (Abrahamsson et al., 2002; Berki, 2004; Sfetsos, Angelis, Stamelos, & Bleris, 2004) mostly report on the advantages of agile methods. Supporters of agile methods, though, must provide convincing answers to questions such as what is the quality of the software produced, or which evidence supports the superiority of agile quality attributes over pure scientific reasoning as this is employed in traditional SE development methods and software tools that support them?

There has been little research work that focuses on agile software development quality issues, while limited industrial practice, often contradicted by the academia, disproves the benefits claimed from agile method(ology) use. Moreover and more often the evidence of the superiority of the agile methods, coming from academia or industry, are contradicting and bound to other uncontrolled experimental factors. Certainly, opinions and observations on agile development have not been derived from formal measurements. No matter how strong the beliefs on the applicability and suitability of agile methods are, these remain subjective claims, and they do not constitute a formal proof of the agile method deployment or, rather, of the quality function deployment.

The XP community, for instance, believes that design information should allow the design to be captured in a form that is machine processable and the components to be reused smoothly integrated. However, software development tools should control the management of this agile development process, enabling the top-down development of applications up to code generation and bottom-up development to include and provide for reengineering of existing codes, fully integrated with the development environment. Furthermore, MetaCASE and CAME tools should be foremost utilised to construct an agile, flexible to the stakeholders and especially to end-users needs, development method.

Societal norms, which are expressed by the value system shared by the majority of a society, have influenced the development of structures and ways of functioning of organisations (Hofstede, 2001). People in a particular organisational setting or software development team share values, beliefs, attitudes, and patterns of behaviour (or not!). The agile approach can be considered to be a culture in its own right aiming at socially-constructed and user-accepted software. The question of course remains open if the socially-constructed part of reality suits for agile method engineering. The literature examined in this chapter has suggested that higher customer involvement also results in higher quality, especially in terms of meeting requirements. Agile methodologies emphasise user satisfaction through user participation, recognition of and response to continuous changing requirements, and frequent delivery of products together with adaptive and iterative software development by self-organising teams, which recognise that team-members are competent professionals, able to choose and follow an adaptive process. A further research question that rises is to which degree agile methods cater for *representative* and *consensus end-user participation* like the traditional ETHICS methodology (see in Mumford, 1983; 2003) caters for? Maximum stakeholder and user involvement in an agile development process does not necessarily mean maximum *end-user* involvement. Otherwise, end-users are not very willing to frequently provide feedback on software artefacts and other devliverables. Hence, usability engineering and cognitive systems engineering principles need to be taken onboard for an agile, holistic design process that encounters human beings, their opinions and feelings.

Frequent delivery of product (incremental), user-involvement, and adaptability to changing or erroneously understood requirements has in recent years been the domain of agile software development. Maintainability, reliability, productivity, and re-engineering issues are all connected to timeliness, which is a demand of agile software development by adopting a frequent product release approach in shorter deliverable cycles. Timeliness and how it is achieved is an issue that has not been adequately researched. Hence, the answer to the question whether agile methods are time-consuming or time-effective remains open for future investigation.

As exposed in this chapter, process and product metrics are useful indicators of improvement. Direct measures of internal characteristics are by far the easiest to obtain. However, external characteristics (such as commitment, job satisfaction, user satisfaction, and knowledge transfer) are complex and often not measurable in simple terms. Measurement activities can only be successful if the development process is mature. This presupposes commitment at all levels within an organisation. Knowledge creation and experience sharing could be formalised and reused in a continuously improving cycle.

A further emerging research question is how will agile methods accommodate the need to keep metrics with the philosophy of minimal documentation? We propose that the use of automated data capture will improve the process and ultimately the product. Future research should probably point to an agile collection of metrics rather than agile metrics, that is the agility will be shifted in the process and not in the actual product metrics, in order to overcome the dangers inherent in the agile philosophy of minimal documentation. These can, perhaps, be addressed by automation tools, efficient ways of data/feedback collection, data/information organisation, organisational memory information systems, procedures for making knowledge explicit, data mining, and requirements changed versions documentation procedures.

Extensions of agile methods constructs, through metamodelling and method engineering principles, to include metrication procedures and be supported by automated tools is also possible in future. The latter could prove particularly fruitful for continuous recording of product metrics to draw comparisons before and after re-factoring. In general, software tools that could support agile methods and agile processes need to be rich in communication and collaboration features in order to realise participative design decisions and requirements management challenges. According to Damian and Zowghi (2003), who report on a field study with cross-functional stakeholder groups, requirements management in multi-site software development organisations, due to increasing globalisation of the software industry, demands to address the following groups of problems: (i) cultural diversity, (ii) inadequate communication, (iii) ineffective knowledge management, and (iv) time zone differences. The previous resulted in inadequate participation, difficulties in requirements' common understanding, delays in deliverables, ineffective decision-making meetings, and a reduced level of trust, to name just but a few, real software development stakeholders' problems.

We propose that in an agile, global software development, situations similar to the previously mentioned problematic circumstances could be supported by suitable groupware, hypertext, e-mail, video-conferencing, or other stakeholder- and end-user-centred information and communication technologies. These process quality-oriented facilities could allow them to inform and be informed on requirements' change, express own opinions, and willingly provide feedback upon frequent information retrieval on the product's development progress. Equally important is to have access to reports on similar success or failure stories on agile development project teams. Such reports will allow a deeper study to the levels of knowledge, communication, and trust that are required to operate in a project team.

According to Holcombe et al. (1998), there is little empirical evidence of the superiority of one method over another and that can be seen clearly in large-scale projects where methodological problems are more obvious. Moreover, there seems to be a crisis of intellectual respectability in the method selection and evaluation subject. Not only the evaluation and quality assurance of the methods under use are weak, the selection of the types of system and problem to focus on (method application domain) restricts the suitability and applicability of any method. In order to convince, in a scientific manner, that a method A is better than a method B in large design projects (where most of the problems really lie), we must present rigorous evidence drawn from carefully controlled experiments of suitable complexity. This is, more or less, impossible in practical terms. Is there an alternative approach? The use of theoretical models of computing systems can provide some alternative approaches (Berki, 2001; Berki, 2006; Holcombe et al., 1998).

The interest for agile methodology designers, therefore, should be in identifying and using general and understandable, groupware-oriented structures that adequately capture the features of changeable specifications, testable computations, collaboration and feedback mechanisms for frequent communication. This can be achieved in terms of specialised and sufficiently general design structures that can capture the richness and testedness of domain specifications, considering at the same time people's cognitive needs (Huotari et al., 1999) and maximum participation and, therefore, empowerment (Berki, 2001). On the other hand, it is important for the various IS stakeholders to state clearly their objectives and expectations from the software products, in order for agile software developers to respond to these characteristics and define the agile final product and agile work processes with features that reflect these required objectives.

Yet, links, opinions, and insights from various related contexts and contents need to be provided for agile software development teams to reach a level of maturity. Notwithstanding the culture of the agile methods paradigm promotes significant working values and exposes scientific knowledge principles in software and IS development that have not been combined and utilised in a similar way before. In order, however, for agile methodologies to finally present an integrated solution based on holistic communication rules within appropriate structures, researchers will have to answer a future research question. That will need to capture the modelling of the semantics, pragmatics, and semiotics of systems' and stakeholders' requirements and thus provide the scientific ground for usability engineering in different cultural contexts.

It is questionable and not, yet, clear if agile and lightweight methods cater for a flexible, lightweight quality or if traditional development methods provide scientific reasoning that is not offered by agile methods. The latter is probably inherent in the nature of agile methodology since it is considered a cooperative game of invention and communication that utilises poetry, Wittgenstein's philosophical concepts, and participative games (see Cockburn, 2002). Cockburn (2002) further defines agile software development as the use of light but sufficient rules of project behaviour and the use of human and communication-oriented rules. On the other hand, "agility" is described as dynamic, context-specific, aggressively change-embracing, and growth-oriented. "It is not about improving efficiency, cutting costs, ... It is about succeeding in emerging competitive arenas, and about winning profits, market share, and customers ..." (Goldman, Nagle, & Preiss, 1995).

In assisting developers to make judgements about the suitability and applicability of agile development methods, Miller and Larson (2006) support that the intentions of an actor are vital to a further *deontological analysis*, while a *utilitarian analysis* forces a broad view of who is affected by an act. Hence, further utilitarian analysis will assist software engineers to think professionally about the consequences for other stakeholders and especially consider the end-users consensus participation, while a deontological viewpoint will always guarantee a "proper," ethical decision.

SUMMARY

This chapter examined and analysed the trends and challenges that exist in traditional software development and in agile software development, in a critical manner. The chapter concluded by committing to the motivation and accomplishment of the agile manifesto initial expectations and ambitions, that is the consideration of a flexible approach in software development by adopting scientific and communication principles for ISD. The belief that these aims can be achieved is emphasised by the suggestion of utilising metamodelling and ME technology. For instance, the generic process architectural constructs of the CDM-FILTERS metamodel encapsulate both communicative and scientific quality properties, which can be utilised in agile method engineering. The latter can automatically be utilised by MetaCASE and CAME tools, which offer an adequate platform that support stakeholder-centred and -originated quality requirements for a flexible construction of agile methods.

This work supports the combination of creative intuition and engineering principles to transform stakeholders' requirements to *adequate* human-centred information systems. Stakeholders' needs will be mirrored successfully in a natural, smooth, and unambiguous manner, if the software artefacts' transformation will be based on agile methods that serve as a communicative platform for understanding, and offer *total quality assurance*.

Considering that commercial applicability with scientific reasoning is likely to increase in ongoing research, the development perspectives for collaborative business values and academic values will, in turn, maximise the likelihood of suitable ISD methods. The ultimate contribution of this analysis could be a critical thinking framework that will provide a dialectical instrument for re-assessing and re-framing the potential applicability and suitability of agile methods construction and adoption. That *metacognitive* and *meta-constructivist* method knowledge itself could give agile, improved ways of work to evaluate and model information systems and people's needs in a scientific and progressive manner.

REFERENCES

Abrahamsson, P. (2002). *The role of commitment in software process improvement.* PhD thesis, University of Oulu, Department of Information Processing Science and Infotech, Finland.

Abrahamsson, P. (2005). Project manager's greetings: Agile greetings. *Agile Newsletter, 1*(1), 2004.

Abrahamsson, P., Salo, O., Ronkainen, J., & Warsta, J. (2002). *Agile software development methods: Review and analysis.* Espoo, Finland: VTT Publications.

Agile COCOMO Web site. (2006). Retrieved May 3, 2006, from http://sunset.usc.edu/cse/pub/research/AgileCOCOMO/

Agile Manifesto Web site. (2005). Retrieved November 27, 2005, from http://www.agilemanifesto.org/

Amber, S. (2005). *Agile documentation: Strategies for agile software development.* Retrieved November 28, 2005, from http://www.agilemodeling.com/essays/agileDocumentation.htm

Avison, D., & Fitzgerald, G. (2003). *Information systems development: Methodologies, techniques, and tools* (3rd ed.). UK: McGraw-Hill Publishing Company.

Avison, D., & Wood-Harper, T. (1990). *Multiview: An exploration in information systems development.* Oxford: Blackwell Scientific Publications.

Beck, K. (2000). *Extreme programming explained: Embrace change.* Reading, MA: Addison Wesley Longman.

Beck, K. (2001). Extreme programming explained. *Software Quality Week*, San Francisco, May.

Beck, K. (2003). *Test driven development: By example.* Boston: Addison Wesley.

Beck, K., & Fowler, M. (2001). *Planning extreme programming.* Boston: Addison Wesley.

Beizer, B. (1995). Foundations of software testing techniques. The *12th International Conference & Exposition on Testing Computer Software*, Washington, DC.

Beizer, B. (1990). *Software testing techniques* (2nd ed.). Van Nostrand Reinhold.

Bell, S., & Wood-Harpet, A. T. T. (1992). *Rapid information systems development: A non-specialist's guide to analysis and design in an imperfect world.* McGraw-Hill.

Berki, E., (2004). Formal metamodelling and agile method engineering in MetaCASE and CAME tool environments. In K. Tigka & P. Kefalas (Eds.), *The 1st South-East European workshop on formal methods.* Thessaloniki: SEERC.

Berki, E. (2006, March). Examining the quality of evaluation frameworks and metamodeling paradigms of information systems development methodologies. In E. Duggan & H. Reichgelt (Eds.), *Measuring information systems delivery quality.* Hershey, PA: Idea Group Publishing.

Berki, E., Georgiadou, E., & Holcombe, M. (2004). Requirements engineering and process modelling in software quality management: Towards a generic process metamodel. *The Software Quality Journal, 12*, 265-283.

Berki, E., Georgiadou, E., & Siakas, K. (1997, March 7-9). A methodology is as strong as the user involvement it supports. *International Symposium on Software Engineering in Universities – ISSEU 97* (pp. 36-51). Rovaniemi.

Berki, E., Isomäki, H., & Jäkälä, M., (2003). Holistic communication modeling: Enhancing human-centred design through empowerment. In D. Harris, V. Duffy, M. Smith, & C. Stephanidis (Eds.), *Cognitive, social, and ergonomic aspects* (Vol. 3 of HCI International) (pp. 1208-1212). Heraklion, Crete: Lawrence Erlbaum Associates.

Berki, E., Lyytinen, K., Georgiadou, E., Holcombe, M., & Yip, J. (2002). Testing, evolution, and implementation issues in metacase and computer assisted method engineering (CAME) environments. In G. King, M. Ross, G. Staples, & T. Twomey (Eds.), *Issues of quality management and process improvement. The 10th International Conference on Software Quality Management,* SQM02. Limerick: Computer Society.

Bock, G. W., Zmud, R. W., Kim, Y. G., & Lee, J. N. (2005). Behavioural intention formation in knowledge sharing: Examining the roles of extrinsic motivators. Socio Psychological Forces and Organisational Climate. *MIS Quarterly, 29*(1), 87-111, March.

Boehm, B. W. (1981). *Software engineering economics.* Prentice Hall.

Boehm, B., & Turner, R. (2003a). Using risk to balance agile and plan-driven methods. *Computer, 36*(6), 57-64.

Boehm, B., & Turner R. (2003b). *Balancing agility and discipline: A guide for the perplexed.* Addison Wesley Professionals.

Burr, A., & Georgiadou, E. (1995). *Software development maturity—a comparison with other industries* (5th ed.). World Congress on Total Quality, India, New Delhi.

Checkland, P. (1981). *Systems thinking, systems practice.* Chichester, UK: Wiley.

Checkland, P., & Scholes J. (1990). *Soft systems methodology in action.* Toronto: John Wiley & Sons.

Chen, A. N. K., & Edgington, T. M. (2005). Assessing value in organizational knowledge creation: considerations for knowledge workers. *MIS Quarterly, 29*(2), 279-309.

Cockburn, A. (2004). *Agile software development.* Cockburn-Highsmith Series. Addison-Wesley.

COCOMO Web site. (2006). Retrieved May 3, 2006, from http://sunset.usc.edu/research/cocomosuite/index.html

Conner, K., & Prahalad, C. (1996). A resource-based theory of the firm: Knowledge versus opportunism. *Organization Science, 7*(5), 477-501.

Couger, J. D., & Adelsberger H. (1988). Comparing motivation of programmers and analysts in different socio/political environments: Austria compared to the United States. *Computer Personnel, 11*(4), 13-17.

Couger, J. D., Borovitz, I., & Zviran, M. (1989). Comparison of motivating environments for programmer/analysts and programmers in the U.S., Israel, and Singapore in Sprague. In R/H. JR (Ed.), *Proceedings of the 22nd Annual Hawaii International Conference on Systems Sciences* (pp. 316-323). Washington, DC: IEEE Computer Society Press.

Couger, J. D., Halttunen, V., & Lyytinen, K. (1991). Evaluating the motivating environment in Finland compared to the United States—a survey. *European Journal of Information Systems, 1*(2), 107-112.

Dalcher, D., & Genus, A. (2003). Avoiding IS/IT implementation failure. *Technology Analysis and Strategic Management, 15*(4), 403-407.

Damian, D. E., & Zowghi, D. (2003). RE challenges in multi-site software development organisations. *Requirements Engineering, 8*(3), 149-160.

DeMarco, T. (1982). *Controlling software projects: Management, measurement, & estimation.* Englewood Cliffs, NJ: Prentice Hall.

DeMarco, T., & Boehm, B. (2002). The agile methods fray. *Computer, 35*(6), 90-92.

Deming, W. E. (1986). *Out of the crisis: Quality, productivity, and competitive position.* MA.

Dorling, A. (1993). Spice: Software process improvement and capability determination. *Software Quality Journal, 2*(93), 209-224.

Dubinsky, Y., Talby, D., Hassan, O., & Keren, A. (2006). *Agile metrics at the Israeli Air force.* Retrieved January 21, 2006, from http://www.cs.huji.ac.il/~davidt/papers/Agile_Metrics_AgileUnited05.pdf

Fenton, N. (1991). *Software metrics—A rigorous approach.* Chapman & Hall.

Fenton, N., & Pfleeger, S. (1997). *Rigorous & practical approach.* PWS Publishing Company.

Fowler, M. (1997). *UML distilled.* Addison Wesley.

Fowler, M. (1997). *Analysis patterns: Reusable object models.* Addison Wesley.

Fowler, M. (1999). *Refactoring: Improving the design of existing code.* Addison-Wesley.

Garud, R., & Kumaraswamy, A. (2005). Vicious and virtuous circles in the management of knowledge: The case of Infosys technologies. *MIS Quarterly, 29*(1), 9-33, March.

George, M. (2003). *Lean six sigma for service: how to use lean speed and six sigma quality to improve services and transactions.* McGraw Hill.

Georgiadou, E. (2003a). GEQUAMO—A Generic, Multilayered, Customisable, Software Quality Model. *Software Quality Management Journal,* December

Georgiadou, E., (2003b). Software Process and Product Improvement: A Historical Perspective. *Cybernetics, and Systems Analysis, 11*(4), 125-142.

Georgiadou, E., & Keramopoulos, E. (2001, April). Measuring the understandability of a graphical query language through a controlled experiment. The *9th International Conference on Software Quality Management, SQM 2001,* University of Loughborough, UK.

Georgiadou, E., & Milankovic-Atkinson, M. (1995, November). Testing and information systems development life cycles. The *3rd European Conference on Software Testing Analysis and Review (EuroSTAR'95),* London.

Georgiadou, E., & Sadler, C. (1995, April). Achieving quality improvement through understanding and evaluating information systems development methodologies. The *3rd International Conference on Software Quality Management, SQM'95,* Seville, Spain.

Georgiadou, E., Hy, T., & Berki, E. (1998). Automated qualitative and quantitative evaluation of software methods and tools. The *12th International Conference of the Israel Society for Quality,* November-December, Jerusalem.

Georgiadou, E., Karakitsos, G., & Sadler, C. (1994b). Improving the program quality by using the re-engineering factor metric ρ. The *10th International Conference of the Israel Society for Quality,* November 1994.

Georgiadou, E., Karakitsos, G., Sadler, C., Stasinopoulos, D., & Jones, R. (1994a). *Program maintainability is a function of structuredness.* Software Quality Management. Edinburg, Scotland: Computational Mechanics Publications, August 1994.

Georgiadou, E., Siakas, K., & Berki, E. (2003). *Quality improvement through the identification of controllable and uncontrollable factors in software development, EuroSPI 2003 (European Software Process Improvement Conference)* (pp. IX-45), Graz, Austria, 10-12.

Geschi, M., Sillitti, A., Succi, G., & Panfilis, G. (2005). Project management in plan-based and agile companies. *IEEE Software, 22*(3), 21-27

Gilb, T. (2006). *Software project management, adding stakeholder metrics to agile projects.* Retrieved January 21, 2006, from http://www.gilb.com/Download/AgileEvo.pdf

Goldman, S., Nagle, R., & Preiss, K. (1995). *Competitors and virtual organisations.* New York: John Wiley & Sons.

Grieves, M. (2005). *Product life cycle management driving the next generation of lean thinking,* McGraw-Hill.

Haase, V. H. (1992, May). Bootstrap: Measuring software management capabilities. First Findings in Europe. *Proceedings of the 4th IFAC/IFIP Workshop,* Austria.

Haase, V., & Messnarz, R. (1994). *B*ootstrap: Fine-tuning process assessment. *IEEE Software, 11*(4), 25-35.

Hennell, M. A. (1991). How to avoid systematic software testing, Software Testing. *Verification Reliability, 1*(1), 23-30.

Herbsleb, J., Carleton, A., Rozum, J., Siegel, J., & Zubrow, D. (1994). *Benefits of CMM-based software process improvement: Initial results.* Technical Report, CMU/SEI-94-TR-13, August.

Hofstede, G. (2001). *Culture's consequences: Comparing values, behaviours, institutions, and organisations*(2nd ed.). Thousand Oaks, CA; London: Sage Publications.

Holcombe, M. (2005, June 18-23). Extreme programming and agile processes in software engineering. The *6th International Conference,* XP 2005, Sheffield, UK (Lecture Notes in Computer Science).

Holcombe, M., & Ipate, F. (1998). *Correct systems: Building a business process solution.* Springer-Verlag.

Holmes M. C. (1995). *The relationship of cross-cultural differences to the values of information systems professionals within the context of systems development.* PhD dissertation, Denton, Texas.

Holz, H., Melnik, G., & Schaaf, M. (2003). Knowledge management for distributed agile processes: models, techniques, and infrastructures. The *1st Workshop on Knowledge Management for Distributed Agile Processes: Models, Techniques, and Infrastructures,* June 9-11, Germany. Retrieved January 25, 2006, from http://www.dwm.unihildesheim.de/homes/schaaf/WETICE03/content.html

Humphrey, W. (1995). *A discipline for software engineering.* Reading, MA: Addison Wesley. Huotari, J., & Kaipala, J. (1999). Review of HCI research—Focus on cognitive aspects and used methods. Proceedings of T. Kakola (Ed.), *IRIS 22 Conference: Enterprise Architectures for Virtual Organisations.* Jyvaskyla, Finland, Keuruselka: Jyvaskyla University Printing House.

Imai, M. (1986). *Kaizen, the key to Japan's competitive success.* The Kaizen Institute.

ISO. (2005). Retrieved November 16, 2005, from http://www.iso.org/

Ives, B., & Järvenpää, S. (1991). Applications of global information technology: Key issues for management. *MIS Quarterly, 15*(1), 33-49, March.

Jackson, M. (1994). Problems, methods, and specialisation. *Software Engineering Journal, 9*(6), 249-255.

Jäkälä, M., & Berki, E. (2004, March 24-26). Exploring the principles of individual and group identity in virtual communities. In P. Commers, P. Isaias, & N. M. Baptista (Eds.), *Proceedings of the 1st IADIS Conference on Web-Based Communities* (pp. 19-26). Lisbon: International Association for the Development of Information Society (IADIS).

Järvinen, J. (1994). On comparing process assessment results: BOOTSTRAP and CMM. *Software Quality Management, SQM94* (pp. 247-261), Edinburgh.

Jayaratna, N. (1994). *Understanding and evaluating methodologies: NIMSAD—A systemic framework.* Berkshire, UK; McGraw Hill.

Johnson, R.E., & Foote, B. (1988). Designing reusable classes. *Journal of OO Programming, 1*(2), 22-35.

Karlström, D., & Runeson, P. (2005). Combining agile methods with stage-gate project management. *IEEE Software, 22*(3), 43-49.

Keen P. W. (1992). Planning globally: Practical strategies for information technology in the transnational firm. In S. Palvia, R. Palvia, & R. Zigli (Eds.), *The global issues of information technology management* (pp. 575-607). Hershey, PA: Idea Group Publishing.

Kelly, S., Lyytinen, K., & Rossi, M. (1996). MetaEdit+: A Fully Configurable Multi-User and Multi-Tool CASE and CAME Environment. In P. Constantopoulos, J. Mylopoulos, & Y. Vassiliou (Eds.), *Advances in Information Systems Engineering, Proceedings of the 8th International Conference CAiSE '96*, pp. 1-21.

Kitchenham, B. (1996). *Software metrics: Measurement for software process improvement.* NCC, Blackwell.

Kitchenham, B., & Pfleeger, S. L. (1996). Software quality: The elusive target. *IEEE Software, 13*(1), 12-21.

Koskinen, M., Liimatainen, K., Berki, E., & Jäkälä, M. (2005, January 2-5). The human context of information systems. In R. H. Sprague, Jr. (Ed.), *The Proceedings of the 38th Hawaii International Conference on Systems Sciences (HICSS 2005)* (pp. 219a-234). Conference CDROM, IEEE Computer Society, IEEE, Los Alamitos, California.

Kumar, K., & Bjorn-Andersen, N. (1990). A cross-cultural comparison of IS designer values. *Communications of the ACM, 33*(5), 528-238, May 1990.

Kuvaja, P. (1999). New developments in software process improvement. *Keynote Speech in Software Quality Management Conference (SQM 99)*. Southampton. Retrieved from http://www.bootstrap-institute.com/assessment.htm

Kuvaja, P., Similä, J., Kranik, L., Bicego, A., Saukkonen, S., & Koch, G. (1994). *Software process assessment and improvement—The BOOTSTRAP approach.* Cambridge, MA: Blackwell Publishers.

Larman, C. (2004). *Agile and iterative development: A manager's guide*. Addison-Wesley.

Luong, F., & Chau, T. (2002). *Knowledge management in agile methods.* Retrieved from sern.ucalgary.ca/courses/SENG/609.24/F2002/slides/KM.ppt

Lyytinen, K., & Hirschheim, R. (1987). Information systems failures—a survey and classification of the empirical literature. Oxford surveys in information technology. In P. I. Zorkoczy (Ed.), *Oxford University Press*, 4, 257-309

Malhotra, Y. (1997, August 15-17). Knowledge management in inquiring organizations. *Proceedings of 3rd Americas Conference on Information Systems (Philosophy of Information Systems Mini-track)* (pp. 293-295). Indianapolis, IN.

Manninen, A., & Berki, E. (2004). Coordinating the quality of requirements change and organisational communication—An evaluation framework for requirements management tools. In D. Edgar-Neville, M. Ross, & G. Staples (Eds.), *New approaches to software quality. Software quality management XII.* University of Kent at Canterbury, British Computer Society.

Marciniak, J. (1994). Encyclopaedia of software engineering. In B. Randell, G. Ringland, & B. Wulf (Eds.), *Software 2000: A view of the future*. ICL and the Commission of European Communities. John Wiley & Sons.

Mellor, S. (2005). Adapting agile approaches to your project needs. *IEEE Software, 22*(3), 17-34, May/June.

Miller, K. W., & Larson, D. (2005). Agile software development: Human values and culture. *IEEE Technology and Society, 24*(4), 36-42.

Moussa, A., & Schware, R. (1992). Informatics in Africa: Lessons from World Bank experience. *World Development, 20*(12), 1737-1752.

Mumford, E. (1983). *Designing human systems for new technology: The ETHICS method.* Retrieved from http://www.enid.u-net.com/C1book1.htm

Mumford, E. (2003). *Redesigning human systems.* Hershey, PA: Idea Group Publishing.

Nonaka, I. (1998). *The knowledge-creating company.* Harvard Business Review on Knowledge Management. Harvard Business School Press.

Nonaka, I., & Takeuchi, H., (1995). *The knowledge creating company: How Japanese companies create the dynamics of innovation.* Oxford University Press.

Palvia, S., Palvia, R., & Zigli, R. (1992). *The global issues of information technology management.* Hershey, PA: Idea Group Publishing.

Paulk, M. C. (1993). Comparing ISO 9001 and capability maturity model for software. *Software Quality Journal, 2*(4), 245-256.

Paulk, M. C. (1995, June 20-22). The rational planning of [software]: Projects. *Proceedings of the First World Congress for Software Quality, ASQC*, San Francisco.

Paulk, M. C., Curtis, B., & Chrissis, M. B. (1993). Capability maturity model (Version 1.1). *IEEE Software, 10*(4), 19-27, July.

Pettichord, P. (2002). *Agile testing, What is it? Can it work?* Retrieved November 28, 2005, from http://www.io.com/~wazmo/papers/agile_testing_20021015.pdf

Pfleeger, L. (1998). *Software engineering, theory, and practice.* Prentice Hall.

Polymenakou, S. (1995, December). Unlocking the secrets of information systems failures: The key role of evaluation. The *5th Greek Conference in Computing,* Athens (pp. 505-519).

Poppendieck, M., & Poppendieck, T. (2003). *Lean software development: An agile toolkit for software development managers.* The Agile Software Development Series.

Pressman, R. (2000). *Software engineering: A practitioner's approach* (European Adaptation). Ch. 19, Fifth Edition.

Rantapuska, T., Siakas, K., Sadler, C., & Mohamed, W. (1999, September 9-11). Quality issues of end-user application development. The *4th International Conference on Software Process Improvement—Research into Education and Training, INSPIRE '99,* Crete, Greece (pp. 77-89).

Robinson, H.M, & Sharp, H. (2003). XP culture: Why the twelve practices both are and are not the most significant things. *Proceedings of the Agile Development Conference (ADC'03).*

Ross, M., & Staples, G. (1995, April). Maintaining quality awareness. The *3rd International Conference on Software Quality Management, SQM95,* Seville (pp. 369-375).

Rossi, M. (1998). *Advanced computer support for method engineering: Implementation of CAME environment in MetaEdit+.* PhD thesis. Jyvaskyla Studies in Computer Science, Economics, and Statistics. Jyvaskyla, Finland: Jyvaskyla University.

Saaksvuori, A., & Immonen, A. (2003). *Product life cycle management.* Springer Verlag.

Sfetsos, P., Angelis, L., Stamelos, I., & Bleris, G. (2004, June). Evaluating the extreme programming system—an empirical study. *The 5th International Conference on Extreme Programming and Agile Processes in Software Engineering,* Garmisch-Partenkirchen, Germany.

Shepperd, M. (1995). *Foundation of software measurement.* Prentice Hall International (UK).

Siakas, K. V. (2002). *SQM-CODE: Software quality management—cultural and organisational diversity evaluation.* PhD thesis, London Metropolitan University.

Siakas, K., & Balstrup, B. (2005, November 9-11). Global software; Sourcing by Virtual Collaboration? *EuroSPI 2005 (European Software Process Improvement and Innovation, Conference),* Budapest, Hungary.

Siakas, K. V., & Georgiadou, E. (2000, September 7-9). A new typology of national and organisational cultures to facilitate software quality management. The *5th International Conference on Software Process Improvement—Research into Education and Training, INSPIRE 2000,* London.

Siakas, K. V., & Georgiadou, E. (2002). Empirical measurement of the effects of cultural diversity on software quality management. *Software Quality Management Journal, 10*(2), 169-180.

Siakas, K., & Georgiadou, E. (2003, April 23-25). The role of commitment for successful software process improvement and software quality management. The *11th Software Quality Management Conference, SQM 2003* (pp. 101-113). Glasgow, UK.

Siakas, K., & Georgiadou, E. (2005). PERFUMES: A scent of product quality characteristics. The *13th Software Quality Management Conference, SQM 2005*, March, Glouchestershire, UK.

Siakas, K. V., & Mitalas, A. (2004). Experiences from the use of the personal software process (PSP) in Greece. Analysis of cultural factors in the *9th International Conference on Software Process Improvement—Research into Education and Training, INSPIRE 2004* (pp. 11-21). Kent, UK.

Siakas, K., Balstrup, B., Georgiadou, E., & Berki, E. (2005b, April). Global software development: The dimension of culture, IADIS (International Association for development of the Information Society). *International Virtual Multi Conference on Computer Science and Information Systems (MCCSIS 2005)—SEA (Software Engineering and Applications).*

Siakas, K. V., Berki, E., & Georgiadou, E. (2003). CODE for SQM: A model for cultural and organisational diversity evaluation. *EuroSPI 2003 (European Software Process Improvement Conference)* (pp. IX, 1-11). Graz, Austria.

Siakas, K., Berki, E., Georgiadou, E., & Sadler, C. (1997). The complete alphabet of quality software systems: conflicts and compromises. The *7th World Congress on Total Quality & Qualex 97* (pp. 603-618). New Delhi, India.

Siakas, K., Georgidaou, E., & Berki, E. (2005a). Agile methodologies and software process improvement, IADIS (International Association for Development of the Information Society). *International Virtual Multi Conference on Computer Science and Information Systems (MCCSIS 2005)—SEA (Software Engineering and Applications).*

Sommerville, I. (2001). *Software engineering* (6th ed.). Pearson Education.

Spiegler, I. (2005). Knowledge management: A new idea or a recycled concept? *Communications of the Association for Information Systems (AIS), 3*(14), 1-23.

Stapleton, J. (1997). *DSDM: A framework for business centred development.* Addison-Wesley.

Tolvanen, J.-P. (1998). *Incremental method engineering with modeling tools.* PhD thesis, Jyvaskyla Studies in Computing, University of Jyvaskyla.

Vitalo, R. L., Butz, F., & Vitalo, J. (2003). *Kaizen desk reference standard.* Lowrey Press.

Zahran, S. (1998). *Software process improvement: Practical guidelines for business success.* Reading, MA: Addison-Wesley.

APPENDIX A: ACRONYMS AND DESCRIPTION

AMDD	Agile Model Driven Development
ASD	Adaptive Software Development
CBD	Component-Based Development
CMM	Capability Maturity Model
DSDM	Dynamic Systems Development Method
ETHICS	Effective Technical and Human Implementation of Computer-Based Work Systems
FDD	Feature Driven Development
IE	Information Engineering
IS	Information Systems
ISD	Information Systems Development
ISDM	Information Systems Development Methodology
IT	Information Technology
ICT	Information and Communications Technology
JSD	Jackson Structured Development
NCC	National Computing Centre
OMT	Object Modelling Technique
SSADM	Structured Systems Analysis and Design Method
SSM	Soft Systems Method
STRADIS	Structured Analysis and Design of Information Systems
TQM	Total Quality Management
VDM	Vienna Development Method
XP	Extreme Programming
Z	Z Specification

APPENDIX B: THE METAEDIT+ TOOL ARCHITECTURE (METAPHOR WEB SITE)

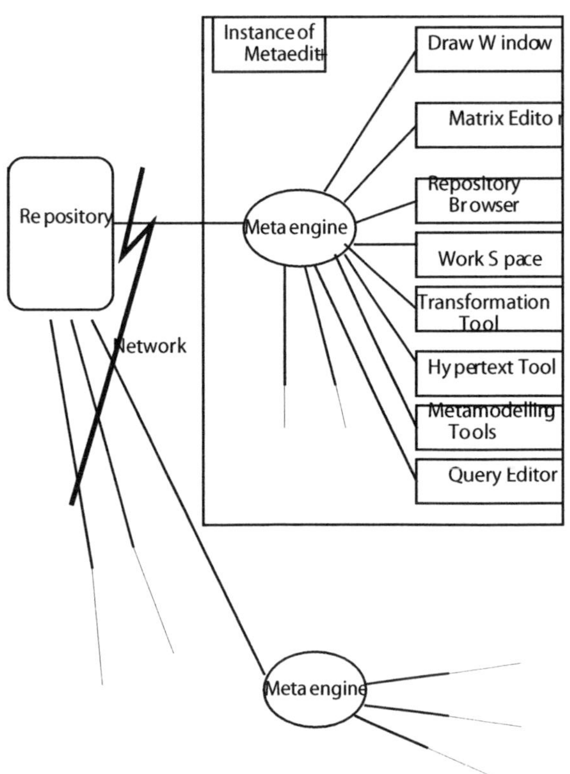

Chapter III
What's Wrong with Agile Methods?
Some Principles and Values to Encourage Quantification

Tom Gilb
Independent Consultant, Norway

Lindsey Brodie
Middlesex University, UK

ABSTRACT

Current agile methods could benefit from using a more quantified approach across the entire implementation process (that is, throughout development, production, and delivery). The main benefits of adopting such an approach include improved communication of the requirements, and better support for feedback and progress tracking. This chapter first discusses the benefits of quantification, then outlines a proposed approach (Planguage), and finally describes an example of its successful use (a case study of the "Confirmit" product within a Norwegian organization, "FIRM").

INTRODUCTION

Agile software methods (Agile Alliance, 2006) have insufficient focus on quantified performance levels (that is, metrics stating the required qualities, resource savings, and workload capacities) of the software being developed. Specifically, there is often no quantification of the main reasons why a project was funded (that is, metrics stating the required business benefits, such as business advancement, better quality of service, and financial savings). This means projects cannot directly control the delivery of benefits to users and stakeholders. In turn, a consequence of this is that projects cannot really control the corresponding costs of getting the main benefits.

Figure 1. A statement made by Lord Kelvin on the importance of measurement (http://zapatopi.net/kelvin/quotes.html)

> *"In physical science the first essential step in the direction of learning any subject is to find principles of numerical reckoning and practicable methods for measuring some quality connected with it. I often say that when you can measure what you are speaking about, and express it in numbers, you know something about it; but when you cannot measure it, when you cannot express it in numbers, your knowledge is of a meagre and unsatisfactory kind; it may be the beginning of knowledge, but you have scarcely in your thoughts advanced to the state of Science, whatever the matter may be."*
>
> Lord Kelvin, 1893

In other words, if you don't *estimate* quantified requirements, then you won't be able to get a realistic *budget* for achieving them. See Figure 1 for a scientist's (Lord Kelvin's) opinion on the need for numerical data!

Further, quantification must be utilized throughout the duration of an agile project, not just to state requirements but to drive design, assess feedback, and track progress. To spell this last point out, *quantification of the requirements* (what do we want to control?) is only a first step in getting control. The next steps, based on this quantification, are *design estimation* (how good do we think our solutions are?) and *measurement of the delivered results* (how good were the solutions in practice?). The key issue here is the active use of quantified data (requirements, design estimates, and feedback) to drive the project design and planning.

One radical conclusion to draw, from this lack of quantification, is that current conventional agile methods are not really suitable for development of industrial products. The rationale for this being that industry is not simply interested in delivered "functionality" alone; they probably already have the necessary business functions at some level. Projects must produce competitive products, which means projects must deliver specific performance levels (including qualities and savings). To address this situation, it is essential that the *explicit* notion of quantification be added to agile concepts.

See Figure 2 for a list of the benefits to agile development of using quantification.

DEFINING QUALITY

The main focus for discussion in this chapter will be the quality characteristics, because that is where most people have problems with quantification. A long held opinion of one of the authors of this chapter (Tom Gilb) is that all qualities are capable of being expressed quantitatively (see Figure 3).

A Planguage definition of "quality" is given in Figure 4. Planguage is a planning language and a set of methods developed by Tom Gilb over the last three decades (Gilb, 2005). This next part of the chapter will outline the Planguage approach to specifying and using quantitative requirements to drive design and determine project progress.

QUANTIFYING REQUIREMENTS

Planguage enables capture of quantitative data (metrics) for performance and resource requirements. A scalar requirement, that is, either a performance or resource requirement, is specified by identifying a relevant scale of measure

Figure 2. What can we do better in agile development (or "at all") if we quantify requirements

Benefits of the Use of Quantification in Agile Development

- Simplify requirements (if the top few requirements are quantified, there is less need for copious documentation as the developers are focused on a clearer, simpler 'message');
- Communicate quality goals much better to all parties (that is, users, customers, project management, developers, testers, and lawyers);
- Contract for results. Pay for results only (not effort expended). Reward teams for results achieved. This is possible as success is now measurable;
- Motivate technical people to focus on real business results;
- Evaluate solutions/designs/architectures against the quantified quality requirements;
- Measure evolutionary project progress towards quality goals and get early & continuous improved estimates for time to completion;
- Collect numeric historical data about designs, processes, organizational structures for future use. Use the data to obtain an understanding of your process efficiency, to bid for funding for improvements and to benchmark against similar organizations!

and stating the current and required levels on that scale. See Figure 5, which is an example of a performance requirement specification. Notice the parameters used to specify the levels on the scale (that is, Past, Goal, and Fail).

EVALUATING DESIGNS

Impact estimation (IE) is the Planguage method for evaluating designs. See Table 1, which shows an example of a simple IE table. The key idea of an IE table is to put the potential design ideas against the quantified requirements and estimate the impact of each design on each of the requirements. If the current level of a requirement is known (its baseline, 0%), and the target level is known (its goal or budget depending on whether a performance requirement (an objective) or a resource requirement respectively, 100%), then the percentage impact of the design in moving towards the performance/resource target can be calculated. Because the values are converted into percentages, then simple arithmetic is possible to calculate the cumulative effect of a design idea (sum of performance and sum of cost) and the performance to cost ratio (see Table 1). You can also sum across the designs (assuming the designs are capable of being implemented together and that their impacts don't cancel each other out) to see how much design you have that is addressing an individual requirement.

Table 1 also shows how you can take into account any uncertainties in your estimates. An additional feature, not shown here, is to assess the credibility of each estimate by assigning a credibility factor between 0.0 and 1.0. Each estimate can then be multiplied by its credibility factor to moderate it.

While such simple arithmetic does not represent the complete picture, it does give a convenient means of quickly identifying the most promising design ideas. Simply filling in an IE table gives a

Figure 3. Tom Gilb's opinion that all qualities can be expressed numerically

The Principle of "Quality Quantification"

All qualities can be expressed quantitatively, "qualitative" does not mean unmeasurable.

Tom Gilb

Figure 4. Planguage definition of "quality"

Definition of Quality

Quality is characterized by these traits:

• A quality describes 'how well' a function is done. Qualities each describe the partial effectiveness of a function (as do all other performance attributes).

• Relevant qualities are either valued to some degree by some stakeholders of the system - or they are not relevant. Stakeholders generally value more quality, especially if the increase is free, or lower cost than the stakeholder-perceived value of the increase.

• Quality attributes can be articulated independently of the particular means (the designs and architectures) used for reaching a specific quality level, even though achievement of all quality levels depend on the particular designs used to achieve quality.

• A particular quality can potentially be a described in terms of a complex concept, consisting of multiple elementary quality concepts, for example, 'Love is a many-splendored thing!'

• Quality is *variable* (along a definable scale of measure: as are all scalar attributes).

• Quality levels are capable of being specified *quantitatively* (as are *all* scalar attributes).

• Quality levels can be *measured* in practice.

• Quality levels can be *traded off* to some degree; with other system attributes valued more by stakeholders.

• Quality can never be perfect (no fault and no cost) in the real world. There are some valued levels of a particular quality that may be outside the state of the art at a defined future time and circumstance. When quality levels increase towards perfection, the resources needed to support those levels tend towards infinity.

(Gilb 2005)

much better understanding of the strengths and weaknesses of the various designs with respect to meeting all the requirements.

Table 1 simply shows estimates for potential design ideas. However, you can also input the actual measurements (feedback) after implementing the design ideas. There are two benefits to this: you learn how good your estimates were for the design ideas implemented, and you learn how much progress you have made towards your target levels. You can then use all the IE table data as a basis to decide what to implement next.

EVOLUTIONARY DELIVERY

The final Planguage method we will discuss is evolutionary project management (Evo). Evo demands include the following:

- That a system is developed in a series of small increments (each increment typically taking between 2% and 5% of the total project timescale to develop).
- That each increment is delivered for real use (maybe as Beta or Field trial) by real "users" (any stakeholder) as early as possible (to obtain business benefits and feedback, as soon as possible).
- That the feedback from implementing the Evo steps is used to decide on the contents of the next Evo step.
- That the highest value Evo steps are delivered earliest, to maximize the business benefit.

Note that "*delivery*" of requirements is the key consideration. Each delivery is done within an Evo step. It may, or may not, include the building or

Figure 5. Example showing Planguage parameters used to specify a performance requirement: "Screen Usability."

Tag: Screen Usability.
Scale: The average number of errors per thousand defined [Transactions] made by system users.
Meter: System to maintain counts for the <different types of error messages> sent to screen.
Past [Order Entry]: 531 ← As measured by Order Entry Department using existing system.
Goal [Order Entry]: < 200 ← Sales Manager.
Fail [Order Entry]: > 400 ← Sales Manager.

Table 1. An example of a simple IE table (Gilb, 2005)

Design Ideas-> Requirements: Goals and Budgets	Idea 1 Impact Estimates	Idea 2 Impact Estimates	Sum for Requirement (Sum of Percentage Impacts)	Sum of Percentage Uncertainty Values	Safety Deviation
Reliability 300 <-> 3000 hours MTBF	1950hr (1650hr) ±0 61%±0	1140hr (840hr) ±240 31%±9%	92%	±9%	-108%
Usability 20 <-> 10 minutes	19min. (1min.) ±4 10%±40%	14min. (6 min.) ±9 60%±90%	70%	±130%	-130%
Maintenance 1.1M <-> 100K/year US$	1.1M $/Y (0 K$/Y) ±180K 0%± 18%	100K S/Y (1 M$/Y) ±720K 100%±72%	100%	±90%	-50%
Sum of Performance	71%	191%			
Capital 0 <-> 1 million US$	500K (500K) ±200K 50%±20	100K (100K) ±200K 10%±20	60%	±40%	-10%
Sum of Costs	50%	10%			
Performance to Cost Ratio	1.42 (71/50)	19.10 (191/10)			

creation of the increment (some Evo steps may simply be further rollout of *existing* software).

Development of necessary components will occur incrementally, and will be continuing in parallel while Evo steps are being delivered to stakeholders. Most development will only start when the decision has been taken to deliver it as the next Evo step. However, there probably will be some increments that have longer lead-times for development, and so their development will need to start early in anticipation of their future use. A project manager should always aim to "buffer" his developers in case of any development problems by having in reserve some components ready for delivery.

Planguage Approach to Change

It is important to note that the quantified requirements, designs, and implementation plans are not "frozen," they must be subject to negotiated change over time. As Beck points out, "Everything in software changes. The requirements change. The design changes. The business changes. The technology changes. The team changes...The problem isn't change, per se,...the problem, rather, is the inability to cope with change when it comes" (Beck, 2000).

Planguage's means of dealing with change are as follows:

- Performance and resource requirements are quantified to allow rapid communication of any changes in levels.
- IE tables allow dynamic reprioritization of design ideas and help track progress towards targets.
- Evo enables all types of change to be catered for "in-flight" as soon as possible. There is regular monitoring of the best next Evo step to take.

DESCRIPTION OF THE PLANGUAGE PROCESS

To summarize and show how the methods (for quantifying requirements, evaluating designs, and evolutionary delivery) described earlier in this chapter fit together, here is a description of the Planguage process for a project:

1. Gather from all the key stakeholders the top few (5 to 20) most critical goals that the project needs to deliver. Give each goal a reference name (a tag).
2. For each goal, define a scale of measure and a "final" goal level. For example:
   - **Reliable:**
   - **Scale:** Mean Time Between Failure.
   - **Goal: 1 month.**
3. Define approximately four budgets for your most limited resources (for example, time, people, money, and equipment).
4. Write up these plans for the goals and budgets (*Try to ensure this is kept to only one page*).
5. Negotiate with the key stakeholders to formally agree the goals and budgets.
6. **Draw up a list of initial design ideas:** Ensure that you decompose the design ideas down into the smallest increments that can be delivered (these are potential Evo steps). Use impact estimation (IE) to evaluate your design ideas' contributions towards meeting the requirements. Look for small increments with large business value. Note any dependencies, and draw up an initial rough Evo plan, which sequences the Evo steps. In practice, decisions about what to deliver in the next Evo step will be made in the light of feedback (that is when the results from the deliveries of the previous Evo steps are known). Plan to deliver some value (that is, progress towards the required goals) in *weekly* (or shorter) increments (Evo steps). Aim to deliver highest possible value as soon as possible.
7. Deliver the project in Evo steps.
   - Report to project sponsors after each Evo step (weekly or shorter) with your best available estimates or measures, for each performance goal, and each resource budget. O*n a single page,* summarize the *progress to date* towards achieving the goals and the costs incurred.
   - Discuss with your project sponsors and stakeholders what design ideas you

should deliver in the next Evo step. This should be done in the light of what has been achieved to date and what is left to do. Maximizing the business benefit should be the main aim.

8. **When all goals are reached:** "Claim success and move on." Free remaining resources for more profitable ventures.

CASE STUDY OF THE "CONFIRMIT" PRODUCT

Tom Gilb and his son, Kai taught the Planguage methods to FIRM (future information research management), a Norwegian organization. Subsequently, FIRM used these methods in the development of their Confirmit product. The results were impressive, so much so that they decided to write up their experiences (Johansen, 2004). In this section, some of the details from this Confirmit product development project are presented.

Use of Planguage Methods

First, 25 quantified requirements were specified, including the target levels. Next, a list of potential design ideas (solutions) was drawn up (see Figure 8 for an example of an initial design idea specification).

The impacts of the potential design ideas on the requirements were then estimated. The most promising design ideas were included in an Evo

Figure 6. Planguage's 10 values for an agile project based around Beck's four values for XP (Beck, 2000, p. 29)

Ten Planguage Values for an Agile Project

Simplicity

1. Focus on real stakeholder values.

Communication

2. Communicate stakeholder values quantitatively.
3. Estimate expected results and costs for weekly steps.

Feedback

4. Generate useful results weekly, to stakeholders, in their environment.
5. Measure all critical aspects of the attempt to generate incremental results.
6. Analyze deviation from initial estimates.

Courage

7. Change plans to reflect weekly learning.
8. Immediately implement valued stakeholder needs, next week.
Don't wait, don't study ('analysis paralysis'), and don't make excuses. Just Do It!
9. Tell stakeholders exactly what you will deliver next week.
10. Use any design, strategy, method, process that works quantitatively well - to get your results. Be a *systems engineer*, not just a programmer. Do not be limited by your 'craft' background in serving your paymasters.

Figure 7. Planguage policy for project management

Planguage Project Management Policy

- The project manager and the project will be judged exclusively on the relationship of progress towards achieving the goals vs. the amounts of the budgets used.
- The project team will do anything legal and ethical to deliver the goal levels within the budgets.
- The team will be paid and rewarded for benefits delivered in relation to cost.
- The team will find their own work process and their own design.
- As experience dictates, the team will be free to suggest to the project sponsors (stakeholders) adjustments to "more realistic levels" of the goals and budgets.

plan, which was presented using an impact estimation (IE) table (see Tables 2 and 3, which show the part of the IE table applying to Evo Step 9. Note these tables also include the actual results after implementation of Evo Step 9). The design ideas were evaluated with respect to "value for clients" vs. "cost of implementation.". The ones with the highest value-to-cost ratio were chosen for implementation in the early Evo steps. Note that value can sometimes be defined by risk removal (that is, implementing a technically challenging solution early can be considered high value if implementation means that the risk is likely to be subsequently better understood). The aim was to deliver improvements to real external stakeholders (customers, users), or at least to internal stakeholders (for example, delivering to internal support people who use the system daily and so can act as "clients").

An IE table was used as a tool for controlling the qualities; estimated figures and actual measurements were input into it.

On a weekly basis:

1. A subset of the quality requirements (the 25 quality requirements defined initially, for delivery after 12 weeks to customers) was selected to work on by one of four parallel teams.
2. The team selected the design ideas they believed would help them reach the quality requirement levels in the next cycle.
3. The team implemented their chosen design ideas and measured the results.
4. The results were input into the IE table. Each next Evo step was then decided based on the results achieved after delivery of the subsequent step.

Note, the impacts described for Confirmit 8.0 (the baseline (0%) "Past" levels) are based on direct customer feedback and internal usability tests, productivity tests, and performance tests carried out at Microsoft Windows ISV laboratory in Redmond USA. The actual results were not actually measured with statistical accuracy by doing a scientifically correct large-scale survey (more intuitive methods were used).

The Results Achieved

Due to the adoption of Evo methods, there were focused improvements in the product quality levels. Table 4 gives some highlights of the 25 final quality levels achieved for Confirmit 8.5. Table 5 gives an overview of the improvements by function (that is, product component) for Confirmit 9.0. No negative impacts are hidden. The targets were largely all achieved on time.

Figure 8. A brief specification of the design idea "Recoding"

Recoding:
Type: Design Idea [Confirmit 8.5].
Description: Make it possible to recode a marketing variable, on the fly, from Reportal.
Estimated effort: four team days.

Table 2. A simplified version of part of the IE table shown in Table 3. It only shows the objective, "productivity" and the resource, "development cost" for Evo Step 9, "recoding" of the marketing research (MR) project. The aim in this table is to show some extra data and some detail of the IE calculations. Notice the separation of the requirement definitions for the objectives and the resources. The Planguage keyed icon "<->" means "from baseline to target level." On implementation, Evo Step 9 alone moved the productivity level to 27 minutes, or 95% of the way to the target level

	EVO STEP 9: DESIGN IDEA: "Recoding"			
	Estimated Scale Level	Estimated % Impact	Actual Scale Level	Actual % Impact
REQUIREMENTS				
Objectives				
Usability.Productivity 65 <-> 25 minutes Past: 65 minutes. Tolerable: 35 minutes. Goal: 25 minutes.	65 – 20 = 45 minutes	50%	65 - 38 = 27 minutes	95%
Resources				
Development Cost 0 <-> 110 days	4 days	3.64%	4 days	3.64%

The customers responded very favorably (see Figure 9).

On the *second* release (Confirmit 9.0) using Planguage, and specifically the Evo method, the Vice President (VP) of marketing proudly named the Evo development method on the FIRM Web site (see Figure 10. A line executive bragging about a development method is somewhat exceptional!).

Details of the quantified improvements were also given to their customers (see Figure 11, which is an extract from the product release for Confirmit 9.0 published on the organization's Web site).

Impact on the Developers

Use of Evo has resulted in increased *motivation* and *enthusiasm* amongst the FIRM developers because it has opened up "*empowered creativity*" *(Trond Johansen, FIRM Project Director).* The developers can now determine their own design ideas and are not subject to being dictated the design ideas by marketing and/or customers who often tend to be amateur technical designers.

Daily, and sometimes more often, product builds, called continuous integration (CI), were introduced. Evo combined with CI, is seen as

Table 3. Details of the real IE table, which was simplified in Table 2. The two requirements expanded in Table 1 are highlighted in bold. The 112.5 % improvement result represents a 20-minute level achieved after the initial 4-day stint (which landed at 27 minutes, 95%) . A few extra hours were used to move from 27 to 20 minutes, rather than use the next weekly cycle.

Current Status	Improvements		Goals			Step 9			
						Design = 'Recoding'			
						Estimated impact		Actual impact	
Units	Units	%	Past	Tolerable	Goal	Units	%	Units	%
			Usability.Replaceability (feature count)						
1.00	1.0	50.0	2	1	0				
			Usability.Speed.New Features Impact (%)						
5.00	5.0	100.0	10	15	5				
10.00	10.0	66.7	20	15	5				
40.00	0.0	0.0	40	30	10				
			Usability.Intuitiveness (%)						
0.00	0.0	0.0	0	60	80				
			Usability.Productivity (minutes)						
20.00	**45.0**	**112.5**	**65**	**35**	**25**	**20.00**	**50.00**	**38.00**	**95.00**
			Development resources						
	101.0	**91.8**	**0**		**110**	**4.00**	**3.64**	**4.00**	**3.64**

Table 4. Improvements to product quality levels in Confirmit 8.5

DESCRIPTION OF REQUIREMENT / WORK TASK	PAST	CURRENT STATUS
Usability.Productivity: Time for the system to generate a survey.	7200 sec	15 sec
Usability.Productivity: Time to set up a typical specified Market Research (MR) report.	65 min	20 min
Usability.Productivity: Time to grant a set of End-users access to a Report set and distribute report login info.	80 min	5 min
Usability.Intuitiveness: The time in minutes it takes a medium experienced programmer to define a complete and correct data transfer definition with Confirmit Web services without any user documentation or any other aid.	15 min	5 min
Workload Capacity.Runtime.Concurrency: Maximum number of simultaneous respondents executing a survey with a click rate of 20 seconds and a response time < 500 milliseconds, given a defined (Survey-Complexity) and a defined (Server Configuration, Typical).	250 users	6000 users

Table 5. Some detailed results by function (product component) for Confirmit 9.0

FUNCTION	PRODUCT QUALITY	DEFINITION (quantification)	CUSTOMER VALUE
Authoring	Intuitiveness	Probability that an inexperienced user can intuitively figure out how to set up a defined simple survey correctly.	Probability increased by 175% (30% to 80%)
Authoring	Productivity	Time in minutes for a defined advanced user with full knowledge of Confirmit 9.0 functionality to set up a defined advanced survey correctly.	Time reduced by 38%
Reportal	Performance	Number of responses a database can contain if the generation of a defined table should be run in 5 seconds.	Number of responses increased by 1400%
Survey Engine	Productivity	Time in minutes to test a defined survey and identify four inserted script errors, starting from when the questionnaire is finished to the time testing is complete and ready for production. (Defined Survey: Complex Survey, 60 questions, comprehensive JScripting.)	Time reduced by 83% and error tracking increased by 25%
Panel Management	Performance	Maximum number of panelists that the system can support without exceeding a defined time for the defined task with all components of the panel system performing acceptably.	Number of panelists increased by 1500%
Panel Management	Scalability	Ability to accomplish a bulk-update of X panelists within a timeframe of Z seconds.	Number of panelists increased by 700%
Panel Management	Intuitiveness	Probability that a defined inexperienced user can intuitively figure out how to do a defined set of tasks correctly.	Probability increased by 130%

Figure 9. An example of pilot customer (Microsoft) feedback

"I just wanted to let you know how appreciative we are of the new 'entire report' export functionality you recently incorporated into the Reportal. It produces a fantastic looking report, and the table of contents is a wonderful feature. It is also a HUGE time saver."

a vehicle for innovation and inspiration. Every week, the developers get their work out onto the test servers and receive feedback.

By May 2005, FIRM had adopted the approach of using a "Green Week" once monthly. In a Green Week, the internal stakeholders are given precedence over the client stakeholders and can choose what product improvements they would like to see implemented. The FIRM developers chose to focus on the evolutionary improvement of about 12 internal stakeholder qualities (such as testability and maintainability).

Initial Difficulties in Implementing Planguage

Even though Planguage was embraced, there were parts of Planguage that were initially difficult to understand and execute at first. These included:

- Defining good requirements ("Scales" of measure) sometimes proved hard (they only had one day training initially, but after the first release saw the value in a week's training!).

Figure 10. Comments by FIRM's VP of marketing, Kjell Øksendal

> "FIRM, through evolutionary development, is able to substantially increase customer value by focusing on key product qualities important for clients and by continuously asking for their feedback throughout the development period. Confirmit is used by the leading market research agencies worldwide and Global 1000 companies, and together, we have defined the future of online surveying and reporting, represented with the Confirmit 9.0."

Figure 11. Confirmit 9.0 release announcement from the FIRM Web site (http://www.firmglobal.com). It gives detail about the method and the quantified product results

> News release
>
> 2004-11-29: Press Release from FIRM
>
> **New version of Confirmit increases user productivity up to 80 percent**
>
> NOVEMBER 29th, 2004: FIRM, the world's leading provider of online survey & reporting software, today announced the release of a new version of Confirmit delivering substantial value to customers including increased user productivity of up to 80 percent.
>
> FIRM is using Evolutionary (EVO) development to ensure the highest focus on customer value through early and continuous feedback from stakeholders. A key component of EVO is measuring the effect new and improved product qualities have on customer value. Increased customer value in Confirmit 9.0 includes:
>
> * Up to 175 percent more intuitive user interface*
> * Up to 80 percent increased user productivity in questionnaire design and testing*
> * Up to 1500 percent increased performance in Reportal and Panel Management*

- It was hard to find "Meters" (that is, ways of measuring numeric qualities, to test the current developing quality levels), which were practical to use, and at the same time measured real product qualities.
- Sometimes it took more than a week to deliver something of value to the client (this was mainly a test synchronization problem they quickly overcame).
- Testing was sometimes "postponed" in order to start the next step. Some of these test postponements were then not in fact done in later testing.

Lessons Learned with Respect to Planguage, Especially the Evo Method

Some of the lessons learned about the use of Planguage, and especially the Evo method, included:

- Planguage places a focus on the measurable product qualities. Defining these clearly and testably requires training and maturity. It is important to *believe* that everything can be measured and to seek guidance if it seems impossible.
- Evo demands dynamic re-prioritization of the next development steps using the ratio of delivering value for clients vs. the cost of implementation. Data to achieve this is supplied by the weekly feedback. The greatest surprise was the power of focusing on these ratios. What seemed important at the start of the project may be replaced by other solutions based on gained knowledge from previous steps.
- An *open architecture* is a pre-requisite for Evo.
- Management support for changing the software development process is another pre-requisite, but this is true of any software process improvement.
- The concept of daily builds, CI, was valuable with respect to delivering a new version of the software every week.
- It is important to control expectations. "Be humble in your promises, but overwhelming in your delivery" is a good maxim to adopt.
- There needed to be increased focus on feedback from clients. The customers willing to dedicate time to providing feedback need identifying. Internal stakeholders (like sales and help desk staff) can give valuable feedback, but some interaction with the actual customers is necessary.
- Demonstrate new functionality automatically with screen recording software or early test plans. This makes it easier for internal and external stakeholders to do early testing.
- Tighter integration between Evo and the test process is necessary.

Conclusion of the Case Study

The positive impacts achieved on the Confirmit product qualities has proved that the Evo process is better suited than the waterfall process (used formerly) to developing the Confirmit product.

Overall, the whole FIRM organization embraced Planguage, especially Evo. The first release, Confirmit 8.5, showed some of Planguage's great potential. By the end of November 2004 with the second release (Confirmit 9.0), there was confirmation that the Evo method can, consistently and repetitively, produce the results needed for a competitive product. Releases 9.5 and 10.0 of Confirmit continued this pattern of successful product improvements delivered to the customers (as of November 2005).

It is expected that the next versions of Confirmit will show even greater maturity in the understanding and execution of Planguage. The plan is to continue to use Planguage (Evo) in the future.

CHAPTER SUMMARY

Use of quantified requirements throughout the implementation of a project can provide many benefits as has been demonstrated by the FIRM organization's use of Planguage (including Evo).

The key messages of this chapter can be summarized in 12 Planguage principles (see Figure 12). By adopting such principles, agile methods would be much better suited for use in the development of industrial products.

Figure 12. Twelve Gilb Planguage principles for project management/software development

Twelve Planguage Principles
1. Control projects by a small set of quantified critical results (that is, not stories, functions, features, use cases, objects, etc.). Aim for them to be stated on one page!
2. Make sure those results are *business* results, not technical.
3. Align your project with your financial sponsor's interests!
4. Identify a set of designs. Ensure you decompose the designs into increments of the smallest possible deliverables.
5. Estimate the impacts of your designs, on *your* quantified goals.
6. Select designs with the best performance to cost ratios; do them first.
7. Decompose the workflow and/or deliveries, into weekly (or 2% of budget) time boxes.
8. Give developers freedom, to find out *how* to deliver those results.
9. Change designs, based on quantified experience of implementation (feedback).
10. Change requirements, based in quantified experience (new inputs).
11. Involve the stakeholders, every week, in setting quantified goals.
12. Involve the stakeholders, every week, in *actually using* increments.

REFERENCES

Agile Alliance. (2006). Retrieved June 2006, from http://www.agilealliance.com/

Beck, K. (2000). *Extreme programming explained: Embrace change.* Addison-Wesley.

Gilb, T. (2005). *Competitive engineering: A handbook for systems engineering. Requirements engineering, and software engineering using Planguage.* Elsevier Butterworth-Heinemann.

Johansen, T. (2004). FIRM: From waterfall to evolutionary development (Evo) or how we rapidly created faster, more user-friendly, and more productive software products for a competitive multi-national market. *Proceedings of European Software Process Improvement (EuroSPI),* Trondheim, Norway, November 10-12, 2004. In T. Dingsøyr (Ed.), Lecture Notes in Computer Science 3281, Springer 2004. See also Proceedings of INCOSE 2005 (Johansen and Gilb 2005) and FIRM Website, http://www.confirmit.com/news/release_20041129_confirmit_9.0_mr.asp/ (Last Accessed: June 2006).

Section II
Quality within Agile Development

Chapter IV
Requirements Specification using User Stories

V. Monochristou
IT Consultant, GNOMON, Informatics, S. A., Greece

M. Vlachopoulou
University of Macedonia, Greece

ABSTRACT

Collecting and analyzing user requirements is undoubtedly a really complicated and often problematic process in software development projects. There are several approaches, which suggest ways of managing user's requirements; some of the most well-known are IEEE 830 software requirements specification (SRS), use cases, interaction design scenarios, etc. Many software experts believe the real user requirements emerge during the development phase. By constantly viewing functional sub-systems of the whole system and participating, in fact, in all phases of system development, customers/users can revise their requirements by adding, deleting, or modifying them. However, in order for this to become possible, it is important to adopt a totally different approach than the traditional one (waterfall model approach), concerning not only the management of user's requirements, but also the entire software development process in general. Agile methodologies represent this different approach since the iterative and incremental way of development they propose includes user requirements revision mechanisms and user active participation throughout the development of the system. The most famous approach concerning requirements specification among the supporters of the agile methodologies is probably user stories. User stories and their main characteristics are thoroughly demonstrated in this chapter. After reading this chapter, the authors hope that the reader may have gained all the basic understanding regarding the use of user stories.

INTRODUCTION

Collecting and analyzing user requirements is undoubtedly a really complicated and often problematic "process" in software development projects. There are several approaches, which suggest ways of managing user's requirements; some of the most well known are IEEE 830 software requirements specification (SRS), use cases, interaction design scenarios, etc.

The success of the final product depends mainly on the success of the previous "process." But how is this "success" defined? Some suggest that the main indicator of success is the compliance of the final product with the initially documented requirements of the customer/user. However, is it plausible to document the real requirements of the customer/user when the entire specification "process" starts and ends before the software development has even started? In addition, after completing this "process," these up-front documented requirements are "locked" without any chance of revision due to the fact that they are often used as the main part of contractual agreements.

A 2001 study performed by M. Thomas (2001) in the UK analyzing 1,027 projects showed that 82% of failed projects report the use of waterfall practices as number one cause of failure, including detailed, up-front requirements. Moreover, a Standish group study presented at XP2002 Conference by Jim Johnson reports that when requirements are specified early in the lifecycle, 45% of features are never used, 19% are rarely used, and 16% are sometimes used.

Since 1986, Parnas and Clements (1986) alleged that it is extremely difficult to write down all the requirements of a system up-front and then to develop it perfectly. Mainly this is because:

- Users and customers do not usually distinguish exactly from the beginning what they want.
- Even if the developers identify all the requirements from the beginning, many of the necessary details will appear during the development of the system.
- Even if all the necessary details could be known up-front, humans are incapable of comprehending so many details.
- Even if humans were capable of comprehending all the details, product and project changes occur.
- People make mistakes.

Many software experts believe that the real user requirements emerge during the development of a system. By constantly viewing functional subsystems of the whole system, and participating in fact in all phases of system development, a user can revise his or her requirements by adding, deleting, or modifying them. However, in order for this to become possible, it is important to adopt a totally different approach than the traditional one (waterfall model approach), concerning not only the management of user's requirements, but also the entire software development process as a whole. Agile methodologies represent this different approach since the iterative and incremental way of development they propose includes user requirements revision mechanisms and user active participation throughout the development of the system.

Rather than making one all-encompassing set of decisions at the outset of a project, we spread the decision-making across the duration of the project. (Cohn, 2004)

User stories is probably the most famous approach concerning requirements specification among the supporters of the agile methodologies. User stories and their main characteristics are thoroughly demonstrated in this chapter.

It has to be stressed that due to their recent appearance, the available bibliography concerning the user stories is not so extensive. One of the most detailed and solid analysis on user stories is Cohn's (2004) *User Stories Applied: For Agile Software Development,* a book that was the inspiration and also the main source of knowledge and understanding for this chapter.

BACKGROUND

The Traditional Approach: Common Pitfalls

"On traditional software development projects we talk about capturing the requirements as if we are engaged in a hunt to cage some wild beasts that lurk in the jungle" (Cohn, 2004). Davies (2005) further alleges that "Requirements are not out there in the project space and waiting to be captured."

A traditional requirements specification document describes the functionalities of a desired software system and acts as both means of communication and data storage. The overall goal of this document is to provide all the necessary information to the development team so as to implement an executable software system, which complies with written user requirements.

The traditional approach (based mainly on the waterfall software life cycle) is to specify the requirements for the whole system up front. That means that both customer/user and the development team have to gain a complete understanding of the final system in the beginning of the development phase. Moreover, these requirements are "locked" and can not be altered during the implementation of the system. In this case, requirements documents are used either as a contract model or as a tool of solving any disputes or misunderstandings.

The idea behind requirements engineering is to get a fully understood picture of the requirements before you begin building the software, get a customer sign-off to these requirements, and then set up procedures that limit requirements changes after the sign-off. (Fowler, 2005)

By using the traditional approach, requirements documents produced are (Davies, 2005):

- **Unidirectional:** Documents are a one-way communication medium. After capturing the requirements, usually through interviews, information flows from author (usually a business analyst) to reader (customer/user). Customer/user has little or no opportunity to feedback (asks questions, contributes ideas and insights). On the other hand, documents may not be as precise as they should be causing misleads to the development team. When working under time-pressure, the development team may have to make its own assumptions on the intended meaning, and this can lead to the development of wrong functionalities in the software system.
- **Selective:** Documents usually include the author's (usually a business analyst) personal perspective on the system. Following traditional development process, it is assumed that the development team does not need to know much about the users' needs and the business environment affecting the system development since requirements are chosen purely on business grounds. The development team is limited to the technical implementation of the system rather than contributing ideas on how to achieve business values.
- **Freezing:** The traditional approach is to specify the requirements for the whole system up-front and then "lock" the requirements. This approach does not take into account the constantly and rapidly altered technological and business environment. As Davis (2005) says, "When we freeze requirements early, we deny the chance to adapt the system to a changing context."

Fowler (2005) mentions the "*unpredictability of requirements,*" According to this, requirements are always changing and this is the norm for three main reasons:

1. **Difficulties on estimation:** Resulting mainly from the fact that software development is a design activity, and thus hard to plan and estimate the cost, but also from the fact that it depends on which individual people are involved. Therefore, it is hard to predict and quantify.
2. **Software's intangible nature:** The real value of a software feature is discovered when the user sees and probably uses early versions of the software. At this time, the customer/user is more likely to understand what features are valuable and which ones are not. It is more than obvious that this "process" can cause changes in the initially agreed requirements.
3. **Business's environment changing nature:** Today, and especially during the last decade, business environments and the fundamental business forces are changing too fast, resulting in constant changes to the software features. "*What might be a good set of requirements now, is not a good set in six months time. Even if the customers can fix their requirements, the business world isn't going to stop for them*" (Fowler, 2005).

Supporters of the agile software development methods strongly believe that development teams should take into consideration the unpredictability of requirements. As predictability of requirements may be a rare reality (with the exception of some projects for organisations [e.g., NASA, Army] where predictability is vital), they insist that a more adaptive approach should be followed.

The Agile Approach

Agile methods are based on quite a different approach that includes:

- Iterative development (small versions of the system with a subset of the required features in short development cycles).
- Customer collaboration (customer in fact participates in all phases of software implementation).
- Constant communication (daily face-to-face communication between project team members).
- Adaptability (last minute changes are allowed).

It is also important to mention that one of the agile manifesto's principles states: "*Welcome changing requirements, even late in development. Agile processes harness change for the customer's competitive advantage.*"

This principle is one of the most important arguments against agile methodologies. The critics insist that something like this is practically impossible to be effective in a software project because any important change that needs to be done in a final stage is simply impossible; and even if it was possible, the total cost and time required would be increased excessively.

From their point of view, "agilists" argue that since the requirements are not crystal-clear at the beginning of any project but they essentially emerge during development, the appropriate "environment" needs to be created so that the acceptance of new or modified requirements (even in final stages) should be possible.

Using agile methods, system features are built iteratively and incrementally. Small versions of the system are continuously presented to the customer/user, so as to use his or her feedback to refine the system in the next iteration. Following this process, "agilists" believe that user requirements emerge during the project.

Iterative development allows the development team to deal with changes in requirements. This presupposes that plans are short term and refer to a single iteration. At the end of each iteration, the team, with the participation of the customer, plans the next iteration and the required features/requirements to be included in that iteration.

The agile requirements approach includes a number of practices that differentiate it from the traditional one. Based on Ambler (2005a), the agile approach embraces among others:

- **Active stakeholder participation:** The stakeholder participates in the requirements process by writing and prioritizing requirements, and by providing domain knowledge and information to the developers. "*Your project stakeholders are the requirements experts*" (Ambler, 2005a).
- **Take a breadth-first approach:** Instead of the big modeling up-front (BMUF) approach, agilists suggest that it is better to attempt to obtain a wider picture of the system in the beginning, trying to quickly gain an overall understanding of the system. Details can be added later when it is appropriate.
- **Lean documentation:** Instead of comprehensive documentation that requires a lot of effort and has doubtable results, agilists consider that a more lean approach, where documentation is as simple as it can possibly be, is more appropriate for software development. In fact, agilists suggest that the development team should create documentation only when it is necessary, and with the precondition that "*the benefit of having documentation is greater than the cost of creating and maintaining it*" (Ambler, 2005b).
- **Small requirements:** Small requirements are much easier to understand, estimate, prioritize, build, and therefore manage.
- **Training on the techniques:** Everyone in the development team, including project stakeholders, should have a basic understanding of the requirements modeling techniques. This implies that stakeholders should gain, primarily, a broad understanding why these techniques are utilized and in which way.
- **Adoption of stakeholder terminology:** Technical terminology and jargon should be avoided as it may prove to be difficult for stakeholders to comprehend. Stakeholders' participation cannot be fully accomplished if their terminology (business terminology) is not used. As Constantine and Lockwood (1999) say, "*avoid geek-speak.*"

At this point, it is very important to mention that applying agile methods in a project for the first time implies a cultural change for most organisations. The adoption of agile methods presupposes a new way of project management and every day business operations for the customer, totally different from the traditional one. Without the genuine support of senior management, active stakeholder participation will likely not be successful, jeopardizing the success of the overall project.

USER STORIES

A Short Description

User stories are one of the primary development artifacts for extreme programming (XP) project teams. They are brief descriptions of features containing just enough information for the developers to produce a reasonable estimate of the effort to implement it.

XP creator Beck (2000) defines a user story as: "*One thing the customer wants the system to do. Stories should be estimable at between one to five ideal programming weeks. Stories should be testable. Stories need to be of a size that you can build a few of them in each iteration*" (Beck & Fowler, 2000).

User stories are unstructured sentences written by the customer with a title and a short paragraph describing the purpose of the story, without technical terms, aiming to define what a proposed system needs to do for them. They focus on user/business needs, goals, and benefits.

An interesting definition is also the one given by Wake (2003): "*A pidgin language is a simplified language, usually used for trade, that allows people who can't communicate in their native language to nonetheless work together. User stories act like this. We don't expect customers or users to view the system the same way that programmers do; stories act as a pidgin language where both sides can agree enough to work together effectively.*"

Another way to view a user story is that it's a reminder for a future conversation between the customer and developers. This conversation takes place usually during iteration planning meeting where the customer/user together with the development team are discussing the details of the stories chosen to be included in the current iteration.

Furthermore, in every user story, one or more acceptance tests are written down, so as to verify that the user story has been correctly implemented.

Using Jeffries terminology (2001), stories in XP have three components: **Cards** (their physical medium), **Conversation** (the discussion surrounding them), and **Confirmation** (tests that verify them).

Typically, the stories are written on 8×13cm paper index cards, although an electronic copy may be used. Lately, some software tools like XPlanner, VersionOne, TargetProcess, Rally, and Select Scope Manager, have appeared, but also in-house tools based on Microsoft Access are referenced. On the other hand, more generic software such as spreadsheets and defect trackers can be used. Some examples of user stories follow:

Story card 2 includes also some indicative acceptance tests.

User Stories and "The Planning Game"

As it is mentioned previously, user stories are used for summarizing user's required features in extreme programming (XP). According to Wake (2002, p. 2), an XP project has three main phases:

1. "*A release planning phase, where the customer writes stories, the programmers estimate them, and the customer chooses the order in which stories will be developed;*

Figure 1. Example story card 1

A doctor can seach for a Patient's Medical Record (PMR) by Name, Personal ID Number, and National Security Number.

Figure 2. Example story card 2

A doctor can edit a PMR only by inserting an extra passwork. • *Try to open a PMR by leaving the password field empty* • *Try to open a PMR by using an invalid password*

2. *An iteration phase, where the customer writes tests and answers questions, while the programmers program; and*
3. *A release phase, where the programmers install the software and the customer approves the result.*"

More specifically, user stories are a central part of the "Planning game," one of the 12 core practices of XP, whose purpose is to rapidly develop a high-level plan for the next release or iteration. The customer and development team cooperate to produce the maximum business value as rapidly as possible. The planning game takes place at various levels (i.e., release planning game, iteration planning game). "*In the release planning game, the goal is to define the set of features required for the next release. It is centered around user stories. The customer writes stories, the programmers estimate the stories, and the customer plans the overall release. The iteration planning game is similar. The customer chooses the stories for the iteration and the programmers estimate and accept the corresponding tasks.*" (Wake, 2002, p. 8)

The basic steps during the planning game are always the same:

- Customer writes user stories to describe the desired features/requirements. It is possible, especially in the beginning, for the development team to help or even to suggest new stories to the customer during an initial story writing workshop. However, "*... responsibility for writing stories resides to the customer and cannot be passed to the developers*" (Cohn, 2004).
- Development team estimates the required effort for each user story to turn into working code in terms of story points. Usually, "*the estimates are in terms of ideal engineering days--days where there are no interruptions, no software or hardware glitches, and the implementation moves smoothly*" (Hayes, 2002). Story points can also be estimated in terms of an ideal hour of work, an ideal week of work, or as a measure of their relative complexity (i.e., story A is 2 times more complex than story B).
- Development team estimates how much effort (story points) can be completed in a single iteration (given time interval). The term velocity is used in this case. It has to be stressed that the team's velocity defines the total productivity (in terms of story points) of a specific development team in a particular iteration, excluding possible overtimes, extra developer's effort, and so on.
- Customer decides which stories to include in the release by prioritizing them. With the necessary information from the developers, customer prioritizes the stories according to the chosen technique. According to the MoSCoW rules technique applied in DSDM, features could be prioritized as:

  - **Must-have** features, which are essential to the system and cannot be left out.
  - **Should-have** features, which are important to the system but the project's success does not rely on these. Due to time constraints some of them could be left out of the release.
  - **Could-have** features, which will be left out if time runs out, without any impact.
  - **Won't have this time** features, which are desired ones that are shifted to the next release.

Furthermore, Beck (2000) suggests that stories can also be sorted by means of:

- "***Value:*** *Business sorts the stories into three piles: (1) those without which the system will not function, (2) those that are less essential*

Figure 3. The planning game

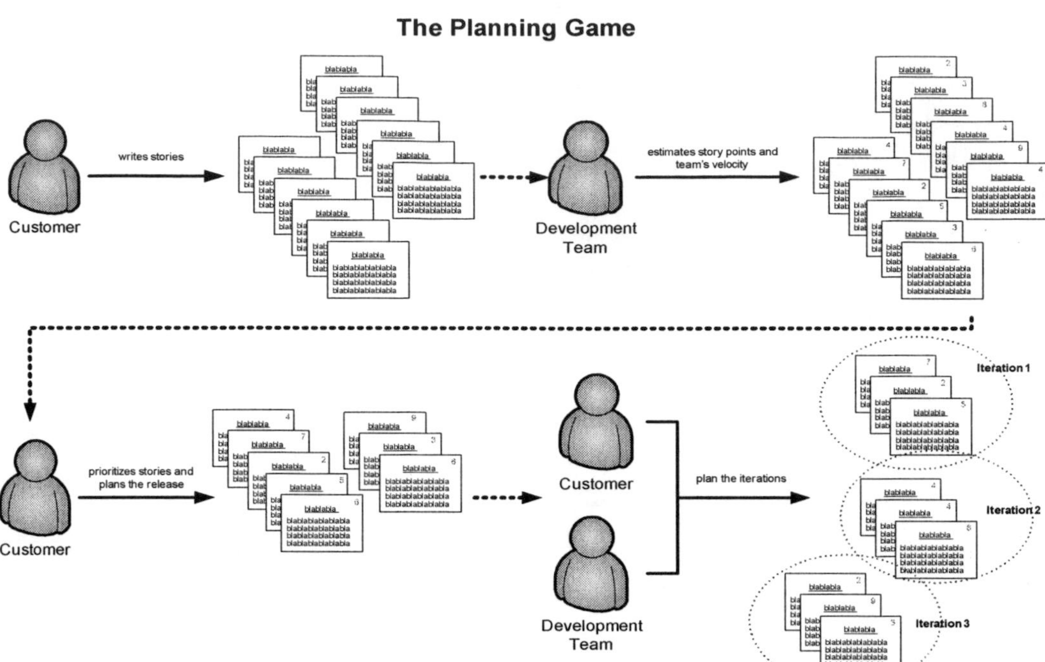

but provide significant business value, and (3) those that would be nice to have.

- ***Risk:*** *Development sorts the stories into three piles: (1) those that they can estimate precisely, (2) those that they can estimate reasonably well, and (3) those that they cannot estimate at all.*"

Independently of the technique, the final goal is to sort the stories in a way that the business value gained will be maximized as much as possible.

- Finally, the whole team (customer and the development team together) plan the current iteration by defining the constituent tasks of each story chosen in this iteration, and by allocating one developer/pair of programmers for each task. At this meeting (the iteration planning meeting), the chosen stories will be discussed in detail and the tasks will be estimated in order to avoid over-allocation.

Although user stories are small by their "nature," "*projects are generally well served by disaggregating them into even smaller tasks.*" This happens firstly because it is more likely that a story will be implemented by more than one developer/pair, and secondly because of the fact that stories "*are not to-do lists for developers*" (Cohn, 2004).

The acronym SMART has been suggested by Wake (2003) for describing the characteristics of good tasks. More detailed, tasks should be:

- **Specific:** It should be clear enough what is involved in a task.
- **Measurable:** Tasks should be measurable in the sense that there are measures "proving" their completion.
- **Achievable:** The developer undertaking a specific task should expect to be able to accomplish it. That means that the developer should have the necessary technical "qualifications" to do the task.

- **Relevant:** Although stories are disaggregating into tasks for the benefit of developers, "*a customer should still be able to expect that every task can be explained and justified*" (Wake, 2003).
- **Time-Boxed:** Tasks should be limited to a specific duration. It is not expected to have an accurate estimation, but it is important to have a rough indication of the duration so that developers may know when they should start to worry.

The following figure shows the way "The planning game" works:

It has to be stressed that during the planning game and before the next iteration, due to the "knowledge" the customer and the team obtain throughout the previous iterations, it is possible that new stories will emerge, whereas some other stories will be modified or even canceled.

Main Attributes for Good Stories

There are six main attributes that a story has to fulfill in order to be characterized as good. Wake (2003) has suggested the acronym INVEST for these six attributes. A good story has to be:

1. **Independent:** User stories should not have interdependencies between them. Otherwise, it is likely to face problems and difficulties not only in prioritization and planning but also in estimation.
2. **Negotiable:** When writing user stories, it has to be kept in mind that they are negotiable. As mentioned previously, user stories are short descriptions of the desired features from the customer's perspective, and not detailed requirements themselves. The details of each story are negotiated in conversations between the customer and the development team during development.
3. **Valuable to users or customers:** As user stories are written by customers/users, they have to describe features that are valuable to them. Stories that are only valued by developers and therefore mainly focused on technological and programming aspects should be avoided. "*Developers may have (legitimate) concerns, but these framed in a way that makes the customer perceive them as important*" (Wake, 2003).

Cohn (2004) points out: "It is very possible that the ideas behind these stories are good ones but they should instead be written so that the benefits to the customers or the user are apparent. This will allow the customer to intelligently prioritize these stories into the development schedule."

4. **Estimatable:** It is essential for stories to be estimatable in a way that developers can estimate the size of the story (in terms of story points). The better the developers estimate the story points the better the customer will prioritize and schedule the story's implementation.

Being not estimatable can be critically affected by:

- The size and the complexity of a story.
- Insufficient domain knowledge.
- Insufficient technical knowledge and team's experience.

5. **Small:** "A good story captures the essence, not the details. Over time, the card may acquire notes, test ideas, and so on, but we don't need these to prioritize or schedule stories" (Wake, 2003).

A story must have the right size in order to be easy to estimate, plan, and prioritize. A small story is more likely to get a more accurate estimation. Stories should represent time between a few man-days of work up to a few man-weeks of work. The usage of index cards, except where the story is likely to include conversation details, notes, and acceptance tests, helps to keep the story small.

There is also an interesting opinion about the size of the title of a good story: "We like to enforce the 5-word rule when writing stories. If your title contains more than five words, then it probably needs refactoring. (...) it is a good exercise to pick out stories that violate this rule and re-examine them." (Industrial Logic, 2004)

6. **Testable:** A good story should be testable. "Writing a story card carries an implicit promise: I understand what I want well enough that I could write a test for it" (Wake, 2003).

Acceptance tests included in the story card should be executed so as to verify that the user story has been correctly implemented. By passing these tests, it is considered that the story has been successfully developed.

Acceptance tests are most useful when automated. Considering the fact that following agile methods, software is developed incrementally, automated tests can help the development team to test constantly and rapidly every change.

Finally, it is normal that there will be some user stories that cannot be automated. For example, "... *a user story that says,* "A novice user is able to complete common workflows without training" can be tested but cannot be automated. Testing this story will likely involve having a human factors expert design a test that involves observation of a random sample of representative novice users" (Cohn, 2004).

User Roles, Personas, and Extreme Characters

Most of the times, requirements are written with the assumption that the system will be used by one generic type of user. What about users that do not belong to the generic type of user?

The use of user roles, personas, or even extreme character technique can prove to be extremely helpful in writing better stories and mainly in minimizing the possibility of omitting important stories.

*A **user role** is a collection of defining attributes that characterize a population of users and their intended interactions with the system." (Cohn, 2004, p. 32). Moreover, "a user role is a particular kind of relationship between some users and a system. (...) A role is an abstraction, a collection of characteristic needs, interests, expectations, and behaviors.* (Constantine & Lockwood, 2006)

Some examples of user roles could be the following:

- **Doctor:** A user that uses the system in order to see past medical data necessary for his prediction, and/or to insert new data after the medical examination has finished.
- **Appointments Secretary:** A user that uses the system in order to reserve doctors' appointments hours, upon patients requests, and to create patient medical record inserting demographical details for first time patients.
- **Nurse:** A user that uses the system in order to see and/or insert data concerning the medical condition, as well as the possible special nutrition and medication instructions given by the doctors.

During the process of identifying the various user roles of a system, it is usual to discover overlaps, interconnections, and relationships among user roles. Sometimes, "*one role may be a specialization of another more general role, one role may be composed of other roles, and roles may otherwise depend on each other.*" (Constantine et al., 2006). In order to avoid this,

Cohn (2004) proposes that the following steps may help a team to identify and select a useful set of user roles:

- **Brainstorming an initial set of user roles:** In a meeting where both the customer and as many developers as possible take part, all the participants start to write down the user roles they picture. Every time a new user role is found, the author informs the whole group. When it becomes difficult to find new roles, the meeting is over. Usually, such a meeting should not exceed 15-20 minutes. Although this meeting may not identify all of the user roles, it is sure that the overwhelming majority are identified.
- **Organizing the initial set:** Once the brainstorming meeting has finished, the group organizes the set of the user roles according to their relationships and their possible overlaps.
- **Consolidating roles:** The user roles, which the group decides, that are of great similarity are consolidated into fewer, more generic ones. At the same time, user roles that have nominal significance to the success of the system are thrown away.
- **Refining roles:** Finally, the group describes each user role by defining its basic attributes. A user role card may ideally include the following attributes:

- "*The frequency with which the user will use the software.*
- *The user's level of expertise with the domain.*
- *The user's general level of proficiency with computers and software.*
- *The user's level of proficiency with the software being developed.*
- *The user's general goal for using the software. Some users are after convenience, others favor a rich experience, and so on.*" (Cohn, 2004, p. 37)

A sample user role card appears in Figure 4.

Cooper (1999) has proposed a variation on user roles called **Personas**. "Rather than abstracting the essential features of a relationship, personas are described as if they were real, specific persons, with personality, detailed history, and complete background." (Constantine et al., 2006).

A persona describes an imaginary named user of the system. The description includes fictitious details such as demographic details and personality characteristics, or even a picture. In order for the use of personas to be helpful for the project, it has to be assured that the personas chosen represent the system's end-users.

A Persona example follows:

Tom is a 45-year-old cardiologist with a lot of experience in his domain. He works in AAA Hospital nine years. He is married and has two daughters. Although he is an early adopter of new practices and novelties in his area of expertise, his familiarity with computers is limited only on Web browsing and e-mail exchanging. However,

Figure 4. A sample user role card

User Role: Doctor

The user will be forced to use the system on a daily basis, in order to extract and insert medical data to Patient Medical Records. He is not quite familiar with the use of computers in general, and up to now he was not utilizing any software related to his occupation. Usefulness and friendliness of the system is important, and no special training should be required.

he is in favor of utilizing the software system being developed, as he recognizes its benefits.

Although the creation of a persona *"can be a fun exercise, the concrete detail can obscure features of the underlying role that are essential for good design"* (Constantine et al., 2006). Moreover, *"there is a risk that using personas to represent the user will not make up for real-life feedback from an actual user. Using personas should, if used, be seen as a complimentary to instead of a substitution of users"* (Toxboe, 2005). Finally, personas' descriptions are often full of unnecessary details that do not contribute anything to the development team and sometimes can also be misleading.

Hence, a combination of using both user roles and personas for selected roles, particularly for some crucial (for the success of the project) roles, may be extremely useful. Following this combination may help the team to better describe the primary user roles and consequently it is more possible to understand better the needs of the system's end-users.

Djajadiningrat, Gaver, and Frens (2000) went a step further. They have proposed the use of **extreme characters.** More specifically, they allege that by considering some extreme characters (e.g., a drug dealer) as possible users of the system can lead to the appearance of stories that would be difficult to picture otherwise.

On the other hand, it is not clear whether these new stories are important to the system, or if the number of the possible end-users "fit" to these stories is significant, in order for their development to be cost-effective.

Techniques for Creating User Stories

Four main techniques have been suggested for creating user stories:

1. **User Interviews:** One of the most widely acceptable techniques for gathering requirements is definitely the user interview. In user stories, the main goal is to interview real users of the proposed system. Alternatively, when this is not possible, user proxies (see next paragraph "User Proxies") replace real users.
   Although most of the people are familiar with interviews and consider them a simple procedure, a really successful interview is a rare reality. The most difficult part is to get users real needs. This can happen only if the questions help the user to express his or her more in-depth opinions, thoughts, and expectations from the system. In fact, the interview should look like a friendly conversation without unnecessary severity. As Cohn (2004, p. 47) alleges, questions in interviews should be *"Open-ended"* and *"Context-free."*
2. **Questionnaires:** Questionnaires may also be used in cases where there is large number of users and is essential to get answers on specific questions. Difficulties arise from the fact that questionnaires exclude the possibility of exchanging information with the user. Therefore, questionnaires are inappropriate for writing new stories, but they can be used for specifying details in existing ones.
3. **Observation:** One additional interesting technique is to observe users' "reactions" and collect their feedback when working with the software. Of course, this technique presupposes that a significant number of real users are available for as long as it takes. If it is possible observation may be extremely helpful, but unfortunately this is not the case all the time. There are project time restrictions and difficulties in accessing real users.
4. **Story-writing workshops:** User stories proponents consider this technique as the most effective way for writing stories. Development teams and users/customers take part in this workshop with the goal to

write as many stories as they can, focusing mainly on quantity and not on quality. According to Cohn (2004, p. 49), "*A good story-writing workshop combines the best elements of brainstorming and low-fidelity prototyping.*" The steps that take place during the workshop can be summarized as following:

- Primary user roles and/or personas are identified.
- Main screen of the system is drawn and basic actions of primary user roles/personas are described.
- Primary stories are generated by the actions described.
- Additional stories are generated by the conversation that follows.

Depending on the project, a combination of the techniques described in this section may be used, resulting in a more effective and comprehensive outcome.

User Proxies

User stories' final quality and effectiveness relies largely on the participation of real users on the customer team. However, on some projects, it is difficult or even impossible to have "on call" real users that will write the stories.

In this case, and according to Cohn (2004), the involvement of **user proxies** could prove to be useful. User proxies are not real users, but for the duration of the project, they could be considered as their representatives making all the necessary decisions. Some indicative types of user proxies, which may be used instead of real users, are following:

- **Users' manager:** In projects that are for internal use instead of access to one or more users, users' manager may be the person that is on-site. It has to be kept in mind that although he may be one of the real users of the system, it is possible that his perception, needs, and priorities from the system may be totally different from his subordinates.
- **IT manager:** Again, in projects for internal use, the IT manager of the customer could be assigned to be the user proxy. Special attention should be given to the fact that often, IT managers consider the introduction of new, sophisticated technology as higher priority.
- **Salespersons/marketing department:** Usually they understand the target market and the real user needs more than anyone else. However, their priorities are mainly affected by the overall "attractiveness" of the system and the consequent sales result. Salespersons are focused on the points they know or assume they will convince the prospective customer/buyer of.
- **Domain experts:** May also be proved very helpful resources, but their usefulness depends on whether their experience is similar to the level of expertise of the system implemented. Their experience may lead to a more complicated system than it was supposed to be.
- **Business/system analysts:** Analysts are frequently used as user proxies and most of the times their involvement could be characterized as successful for the project because "*they have one foot in the technology world and one foot in the domain of the software*" (Cohn, 2004, p. 61). However, some analysts believe their experience and knowledge are quite enough, and therefore conversation with users could be minimized. It is more than obvious that this perception may lead to wrong assumptions.

Conclusively, although the participation of real users may be ideal for writing good user stories, when this is impossible, the use of user proxies, under some conditions, may also lead to the same result. More specifically, because of the shortcomings that each user proxy has, using the right combination (depending on the project and the target users) of user proxies may eliminate these shortcomings.

On the other hand, the early release of a beta version may open an additional communication path to real users, helping to "discover" the difference between real users' perception and the initial user proxies' view of the system.

Acceptance Testing

In user stories, acceptance tests are part of the story. More specifically, tests are written in the story card before the coding starts.

Ideally, tests are specified by customer/users. "*The customers write tests story-by-story. The question they need to ask themselves is, "What would have to be checked before I would be confident that this story was done?*" (Beck, 2000). However, most of the times, it is hard for a customer to write acceptance tests without the assistance of a programmer or a tester. "*That's why an XP team of any size carries at least one dedicated tester. The tester's job is to translate the sometimes vague testing ideas of the customer into real, automatic, isolated tests.*" (Beck, 2000).

Tests are usually written as details when customers and developers discuss a story for the first time, but mainly at the iteration planning meeting (before the iteration starts) when stories are discussed more explicitly. At this point, developers may also write some additional stories in cases where they estimate by experience that the implementation may be complicated and additional points should be tested.

As the customer writes tests, it is more than possible that tests will cover only functional aspects of the system. Although these tests are crucial for the success of the system, there are also some aspects of the system that developers have to test. According to Cohn (2004, p. 72), other important types of testing, which have to be considered are usability testing, performance testing, user interface testing, stress testing, etc.

Because of their large number and the necessity for constant execution, agilists propose that acceptance tests should be automated as much as possible. Two of the most popular tools are FIT and FitNesse.

User Stories and Quality

Quality assurance in the requirements management process defined as the compliance with a quality standard such as ISO, TickIT, etc., is one of the central critique points against user stories. In order to be more precise, quality assurance is one of the most controversial issues when utilizing agile methods.

As it was mentioned earlier in this chapter, agilists believe that requirements emerge during the development phase of the system, therefore, the overall quality of the final system is critically dependent on the challenging task of uncovering the "real" customer requirements. User stories, according to their supporters, provide the appropriate framework toward this direction, contributing thereby in the successful delivery of a final system meeting nearly to the fullest the customer needs. This way, high levels of customer satisfaction are achieved, which subsequently is accounted by many as the number one quality criterion.

In accordance to this, it is important to mention that the Institute of Electrical and Electronic Engineers (IEEE) defines quality in a software system as "*the degree to which a system, component or process meets customer or user needs or expectations.*"

Extreme programming (XP) considers quality of the software produced to be the highest priority. This is accomplished through a series of continuous testing at two basic levels:

1. At a first level, quality of the code is maximized through **test driven development** and **unit testing**. "*Each class implemented*

must have programmer-developed unit tests, for everything that "could possibly break." These tests are to be written during coding of the class, preferably right before implementing a given feature. Tests are run as frequently as possible during development, and all unit tests in the entire system must be running at 100% before any developer releases his code" (Jeffries, 2006).

2. At a second level of testing called **functional testing**, the quality of the system from the business perspective is attempted to be assured. This is achieved through the constant customer/user participation who writes the user stories and respectively acceptance tests with the help of the development team. Customer/user is the one that verifies their accomplishment. According to Crispin (2001), "*If the customer is clear about his acceptance criteria and these are reflected accurately in the acceptance tests, we're much more likely to achieve the level of quality the customer wants.*" Moreover, an acceptance test "*helps verify that a story works correctly by documenting what inputs are supplied to a system and what outputs are expected*" (Reppert, 2004).

In "Extreme Programming Explained," Beck (2000) describes these two levels of testing as internal and external quality. "*External quality is quality as measured by the customer. Internal quality is quality as measured by the programmers.*" In fact, by using acceptance tests, the external quality of the system is likely to be maximized.

Finally, Crispin's (2001) view is an illustrative example of how quality is treated in XP, and generally in agile methods: "*When I started working on XP projects, I realized it wasn't about MY quality standards—it was the customers.*"

Benefits and Limitations of User Stories

There are many positive aspects concerning user stories in contrast with the traditional approaches. Some of the user stories' advantages are mentioned next:

- They favor face-to-face communication instead of written requirements. Constant conversations between the development team and the customer/users may help to **overcome written language inaccuracies**, and help to maintain a close relationship with the customer. This close relationship results in **more effective knowledge transfer** across the team and facilitates the **active participation of the users in designing** the behavior of the system. In this case, the term "*participatory design*" is used (Kuhn & Muller, 1993; Schuler & Namioka, 1993) in contrast to the "traditional" approach where users are not part of the team and all the decisions on the design of the system are made by designers who only study users' written requirements.
- Since stories are written by users, they **are more comprehensible** to them than any other technique. Moreover, their small size, their persistence to simplicity, and the avoidance of technical jargon enforce the previous characteristic.
- They are **suitable for iterative development**. "*Stories work well for iterative development because of how easy it is to iterate over the stories themselves. (...) I can write some stories, code, and test those stories, and then repeat as often as necessary.*" (Cohn, 2004, p. 149). Stories may be written and revised while the development takes place and knowledge of the team becomes more specific.

On the other hand, some of the limitations of user stories could be summarized as follows:

- As it is mentioned previously, traditional requirement documents are also used as a contract model signed usually before the development phase starts, and as a tool of solving any disputes or misunderstandings that may emerge during the development. Although contracts do not assure project success, managers are accustomed to this since it is presumed that contracts can secure them against any future problems. Therefore, it is difficult for them to accept a less "secure" approach where user's requirements do not constitute the technical annex of a contract.
- There are cases where requirements traceability is obligatory because of internal Quality Systems (e.g., ISO 9001, TickIT). This may require additional documentation and a more bureaucratic "attitude" that comes in sharp contrast with the agile approach. However, agilists insist that it is possible to reach the golden mean by adopting a lightweight solution that fulfils basic quality system's documentation requirements but also enables agility.
- Although customer participation throughout the project decreases the risk of system to end a failure, in some cases it may be difficult or it may be a significant cost for the customer.
- On large projects with large teams, communication problems may occur, since face to face conversations and meetings with the participation of the whole team may be hard to organize on a constant basis.

User Stories in Comparison

Apart from user stories, the three more common approaches to user requirements are IEEE 830 software requirements specification (SRS), use cases, and interaction design scenarios. The main differences between user stories and these approaches are described briefly in the following paragraphs:

- **IEEE 830**-style approach implies that all the requirements of the system will be written by analysts (with the assistance of some users) who will "imagine" the planned system even before the development starts. This approach does not take into consideration the benefit of users' feedback once they see part or an early version of the system. Every change in the initially agreed requirements is con sidered as "change of scope" and therefore should be avoided or in the worst case minimized.

Moreover, as customers/users hardly participate actively in requirements specification (their participation is usually limited to few interviews), it is common that the requirements agreed do not entirely cover user's/business' goals.

Finally, requirements documents "produced" are usually extremely large with many details. Producing such documents is not only time consuming but it is likely that customers/users may not understand all the details included in the document.

- According to Cohn (2004, p. 140) "***Use cases** are written in a format acceptable to both customers and developers so that each may read and agree to use cases. The purpose of the use case is to **document an agreement** between the customer and the development team. Stories, on the other hand, are written to facilitate release and iteration planning, and to **serve as placeholders for conversations** about the users' detailed needs.*"

Another important difference is in their level of detail. As stories are written on index cards, their size is limited and therefore their development time is easier to be estimated (by definition the implementation of a story should be competed in a single iteration). In contrast, use cases almost always cover a much larger scope and their implementation is independent of any time considerations.

Moreover, documentation "produced" during use cases' specification is kept after software release to aid in software maintenance. On the other hand, user stories are not preserved as software documentation and are often discarded after the project ends.

Finally, it is usual for use cases to include user interface details that could lead to preconceptions in an early stage.

- Like the previous approaches, **interaction design scenarios** contain much more details than user stories, focusing mainly on describing the personas (see section "User Roles, Personas, and Extreme Characters") and their interaction with the system. Moreover, they usually cover a much larger scope than stories. In fact, most of the times, one scenario may be equal to many stories.

FUTURE TRENDS

As it is cited many times in this chapter, user stories are one of the primary development artifacts for extreme programming (XP) project teams. Bibliography was used to bound user stories with extreme programming since they originated as part of XP. Today, many authors suggest that user stories could be used also with other agile methods. Cohn for example, in "*User Stories Applied: For Agile Software Development*" dedicates a whole chapter describing how user stories can be used effectively with SCRUM.

Moreover, the use of extreme programming is constantly growing while the adoption level of agile methodologies in general is rapidly increased. A few data from research supporting this aspect are distinctively cited:

- The Software Development Times Magazine (July 2003 Issue) published the research results of an Evans Data Corporation research, according to which a rate of 9% of North America companies totally use XP method in their projects.
- A Cutter Consortium research (Charette, 2001) conducted among 200 managing directors/IT managers constituting a respective sample from the point of the geographical allocation of the companies' type and size, recorded that during 2001, 21% of the participants used agile methodologies to more than 50% of their projects. Additionally, in the year 2002, 34%, and in 2003, almost half of the participants expected that more than 50% of their projects would be conducted using agile methodologies.
- The research of Giga Information Group (Sliwa, 2002) in 2002 anticipated that in a period of the following 18 months, 2/3 of IT companies in the U.S. would mostly use agile methods in their projects.

As most of the "agilists" strongly recommend user stories as the appropriate technique for gathering user requirements, it is more than possible that the rise of agile methods will drift the adoption of user stories.

CONCLUSION

User stories are not just another technique on managing user requirements. User stories as part of agile methods propose a totally different approach based on customer active participation throughout the implementation of a system. A customer, with the help of the development team, writes the requirements in his "language," prioritizes them, and finally writes the acceptance tests. "Agilists" believe that this constant customer participation increases drastically the possibility of delivering a system closer to the real user requirements.

However, despite the many positive and innovative ("revolutionary" for some) aspects of the user stories, development teams globally are still trusting and using traditional approaches with comprehensive documentation arguing about the effectiveness of user stories. The main reason behind this skepticism is the fact that, usually, written user's requirements represent the subject of signed contracts. These contracts operate as a means of solving any disputes or misunderstandings between the parties, and their absence frightens mainly the top management. In addition to this, the adoption of user stories and agile methods principles in general, require a change on the "company's culture," something difficult to be accepted by large, bureaucratic organisations.

But, as in one of the four agile manifesto's values quoted, "*People and interactions over processes and tools,*" what is more important is the genuine, honest, and constant cooperation with the customer; this targets to the development of relations of trust rather than the exact and absolute compliance with the terms and conditions of a contract which, even though it is necessary, does not presuppose the success of a project.

Furthermore, in accordance with another agile manifesto's value, "*Responding to change over following a plan,*" user stories and iterative development allow the development team to accept changes in users requirements derived from experiences gained during development.

Conclusively, although user stories appearance is very recent, and their use is not so extensive, it looks like they can solve many of the limitations of the traditional approaches. And as user requirements specification is considered as one of the main factors of a project success, it is possible that using user stories may enhance the success rate of software projects in general. Nevertheless, before adopting a more agile approach based on user stories, it has to be taken into account that user stories have also limitations. "*Due to the need for direct communication, XP is only viable for small co-located teams with access to an onsite customer. Large distributed teams may need to rely on more documentation and adopt RUP or other less agile processes.*" (Davies, 2001)

REFERENCES

Ambler, S. (2005a). *Agile requirements best practices.* Retrieved December 28, 2005, from http://www.agilemodeling.com/essays/agileRequirementsBestPractices.htm

Ambler, S. (2005b). *Agile documentation: strategies for agile software development.* Retrieved December 28, 2005, from http://www.agilemodeling/com/essays/agileDocumentation.htm

Beck, K. (2000). *Extreme programming explained; embrace change.* Reading, MA: Addison Wesley.

Beck, K., & Fowler, M. (2000b). *Planning extreme programming.* Reading, MA: Addison Wesley.

Charette, R. (2001). The decision is in agile versus heavy methodologies. *Cutter Consortium, e-Project Management Advisory Service, Executive Update, 2*(19), 1-3.

Cohn, M. (2004). *User stories applied: For agile software development.* Reading, MA: Addison-Wesley.

Constantine, L. L., & Lockwood L. A. (1999). *Software for use: A practical guide to the models and methods of usage-centered design.* Reading, MA: Addison-Wesley.

Constantine L. L., & Lockwood Ltd Website (2006). *Frequently asked questions.* Retrieved January 18, 2006, from http://www.foruse.com/questions/index.htm#5

Cooper, A. (1999). *The inmates are running the asylum*. Indianapolis: SAMS.

Crispin, L. (2001). *Is quality negotiable?* Retrieved June 28, 2003, from www.xpuniverse.com/2001/xpuPapers.htm

Davies, R. (2001). *The power of stories*. Presented at XP2001 Conference, Sardinia.

Davies, R. (2005). Agile requirements. *Methods & Tools, 13*(3), 24-30, Fall 2005.

Djajadiningrat, J. P., Gaver, W. W., & Frens, J. W. (2000). Interaction relabelling and extreme characters: Methods for exploring aesthetic interactions. *Symposium on Designing Interactive Systems 2000* (pp. 66-71).

Fowler, M. (2005). *The new methodology*. Retrieved December 29, 2005, from http://martinfowler.com/articles/newMethodology.html

Hayes, S. (2002). The problems of predictive development. *Methods & Tools, 10*(4), 22-26, Winter, 2002.

Jeffries, R. (2001). *Essential XP: Card, conversation, confirmation*. Retrieved December 29, 2005, from http://www.xprogramming.com/xpmag/EXPCardConversationConfirmation.htm

Jeffries, R. (2006). XP *questions and answers—quality assurance*. Retrieved April 12, 2006, from http://www.xprogramming.com/qa/xp_q_and_a_QA.htm

Industrial Logic Inc. Web site. (2004). *Storytelling*. Retrieved January 17, 2006, from http://www.industrialxp.org/storytelling.html

Kuhn, S., & Muller, M. (1993). Introduction to the special section on participatory design. *Communications of the ACM, 36*(6), 24-28.

Parnas, D. L., & Clements, P. C. (1986). A rational design process: How and why to fake it. *IEEE Transactions on Software Engineering, 12*(2), 251-257.

Reppert, T. (2004). Don't just break software make software. *Better Software*, 18-23.

Schuler, D., & Namioka, A. (1993). *Participatory design: Principles and practices*. Hillsdale, NJ: Erlbaum.

Sliwa, C. (2002). Users warm up to agile programming. *Computer World, 36*(12), 8.

Stapleton, J. (2003). *DSDM: Business focused development*. Reading, MA: Addison-Wesley.

Thomas, M. (2001). *The modest software engineer*. Retrieved January 29, 2006, from www.safety-club.org.uk/resources/164/MartynThomas.pdf

Toxboe, A. (2005). *Introducing user-centered design to eXtreme programming*. Copenhagen Business School, Department of Informatics.

Wake, W. C. (2002). *Extreme programming explored*. Reading, MA: Addison Wesley.

Wake, W. C. (2003). *INVEST in good stories and SMART tasks*. Retrieved December 29, 2005, from http://xp123.com/xplor/xp0308/index.shtml

Chapter V
Handling of Software Quality Defects in Agile Software Development

Jörg Rech
Fraunhofer Institute for Experiemental Software Engineering (IESE), Germany

ABSTRACT

Software quality assurance is concerned with the efficient and effective development of large, reliable, and high-quality software systems. In agile software development and maintenance, refactoring is an important phase for the continuous improvement of a software system by removing quality defects like code smells. As time is a crucial factor in agile development, not all quality defects can be removed in one refactoring phase (especially in one iteration). Documentation of quality defects that are found during automated or manual discovery activities (e.g., pair programming) is necessary to avoid wasting time by rediscovering them in later phases. Unfortunately, the documentation and handling of existing quality defects and refactoring activities is a common problem in software maintenance. To recall the rationales why changes were carried out, information has to be extracted from either proprietary documentations or software versioning systems. In this chapter, we describe a process for the recurring and sustainable discovery, handling, and treatment of quality defects in software systems. An annotation language is presented that is used to store information about quality defects found in source code and that represents the defect and treatment history of a part of a software system. The process and annotation language can not only be used to support quality defect discovery processes, but is also applicable in testing and inspection processes.

INTRODUCTION

The success of software organizations—especially those that apply agile methods—depends on their ability to facilitate continuous improvement of their products in order to reduce cost, effort, and time-to-market, but also to restrain the ever increasing complexity and size of software systems. Nowadays, industrial software development is a highly dynamic and complex activity, which is not only determined by the choice of the right technologies and methodologies, but also by the knowledge and skills of the people involved. This increases the need for software organizations to develop or rework existing systems with high quality within short periods of time using automated techniques to support developers, testers, and maintainers during their work.

Agile software development methods were invented to minimize the risk of developing low-quality software systems with rigid process-based methods. They impose as little overhead as possible in order to develop software as fast as possible and with continuous feedback from the customers. These methods (and especially extreme programming (XP)) are based upon several core practices, such as *simple design*, meaning that systems should be built as simply as possible and complexity should be removed, if at all possible.

In agile software development, organizations use quality assurance activities like refactoring to tackle defects that reduce software quality. *Refactoring* is necessary to remove *quality defects* (i.e., bad smells in code, architecture smells, anti-patterns, design flaws, negative design characteristics, software anomalies, etc.), which are introduced by quick and often unsystematic development. As time is a crucial factor in agile development, not all quality defects can be removed in one refactoring phase (especially in one iteration). But the effort for the manual discovery, handling, and treatment of these quality defects results in either incomplete or costly refactoring phases.

A common problem in software maintenance is the lack of documentation to store this knowledge required for carrying out the maintenance tasks. While software systems evolve over time, their transformation is either recorded explicitly in a documentation or implicitly through a versioning system. Typically, problems encountered or decisions made during the development phases get lost and have to be rediscovered in later maintenance phases. Both expected and unexpected CAPP (corrective, adaptive, preventive, or perfective) activities use and produce important information, which is not systematically recorded during the evolution of a system. As a result, maintenance becomes unnecessarily hard and the only countermeasures are, for example, to document every problem, incident, or decision in a documentation system like bugzilla (Serrano & Ciordia, 2005). The direct documentation of quality defects that are found during automated or manual discovery activities (e.g., code analyses, pair programming, or inspections) is necessary to avoid wasting time by rediscovering them in later phases.

In order to support software maintainers in their work, we need a central and persistent point (i.e., across the product's life cycle) where necessary information is stored. To address this issue, we introduce our annotation language, which can be used to record information about quality characteristics and defects found in source code, and which represents the defect and treatment history of a part of a software system. The annotation language can not only be used to support quality defect discovery processes, but is also applicable for testing and inspection processes. Furthermore, the annotation language can be exploited for tool support, with the tool keeping track and guiding the developer through the maintenance procedure.

Our research is concerned with the development of techniques for the discovery of quality defects as well as a quality-driven and experience-based method for the refactoring of large-scale software systems. The instruments developed consist of a technology and methodology to support decisions of both managers and engineers. This support includes information about where, when, and in what configuration quality defects should be engaged to reach a specific configuration of quality goals (e.g., improve maintainability or reusability). Information from the diagnosis of quality defects supports maintainers in selecting countermeasures and acts as a source for initiating preventive measures (e.g., software inspections).

This chapter targets the handling of quality defects in object-oriented software systems and services. It is concerned with the theory, methodology, and technology for the handling of defects that deteriorate software qualities as defined in ISO 9126 (e.g., maintainability, reusability, or performance). We describe the relevant background and related work concerning quality defects and quality defect handling in agile software projects, as well as existing handling techniques and annotation languages. The subsequent section encompasses the morphology of quality defects as well as their discovery techniques. As the core of this chapter, we present the techniques for handling quality defects after their discovery in an agile and time-critical environment and define an annotation language to record information about quality defects and their history in source code. Thereafter, a section is used to describe the annotation language that is used to record the treatment history and decisions in the code itself. Finally, we summarize several lessons learned and requirements one should keep in mind when building and using quality defect handling methods and notations in an agile environment. At the end of this chapter, we summarize the described approach and give an outlook to future work and trends.

BACKGROUND

This section is concerned with the background and related work in agile software engineering, refactoring, and quality defects. It gives an overview of quality defect discovery, the documentation of defects, as well as source code annotation languages.

Agile Software Development

Agile software development methods impose as little overhead as possible in order to develop software as fast as possible and with continuous feedback from the customers. Agile methods have

Figure 1. Agile software development (here the XP process)

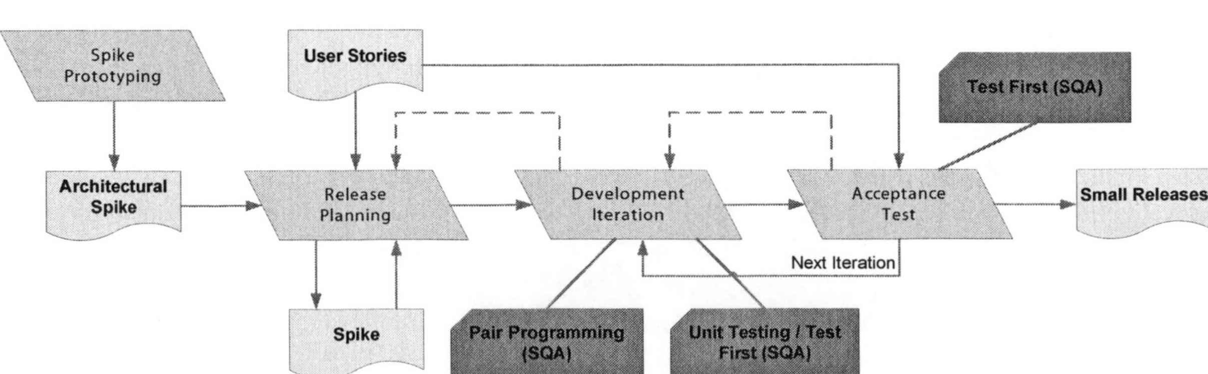

in common that small releases of the software system are developed in short iterations in order to create a running system with a subset of the functionality needed for the customer. Therefore, the development phase is split into several activities, which are followed by small maintenance phases. In contrast to traditional, process-oriented SE, where all requirements and use cases are elicited, agile methods focus on few essential requirements and incrementally develop a functional system in several short development iterations.

Today, extreme programming (XP) (Beck, 1999) is the best-known agile software development approach. 0 shows the general process model of XP, which is closely connected to refactoring, basically being its cradle (Beck & Fowler, 1999).

These agile methods (and especially extreme programming (XP)) are based upon 12 principles (Beck, 1999). We mention four of these principles, as they are relevant to our work.

1. **Planning Game** is the collective planning of releases and iterations in the agile development process and is necessary for quickly determining the scope of the next release. If the requirements for the next iteration are coherent and concise, more focus can be given to one topic or subsystem without making changes across the whole system.
2. **Small releases** are used to develop a large system by first putting a simple system into production and then releasing new versions in short cycles. The smaller the change to the system, the smaller the risk of introducing complexity or defects that are overlooked in the refactoring (or SQA) phases.
3. **Simple design** means that systems are built as simply as possible, and complexity in the software system is removed, if at all possible. The more understandable, analyzable, and changeable a system is, the less functionality has to be refactored or reimplemented in subsequent iterations or maintenance projects.
4. **Refactoring** is necessary for removing qualitative defects that are introduced by quick and often unsystematic development. Decision support during refactoring helps the software engineer to improve the system.

In the highly dynamic processes used in agile methods, teams and organizations need automated tools and techniques that support their work without consuming much time. Especially in the refactoring phase, where the software is revised, automation can be used to detect *quality defects* such as code smells (Fowler, 1999), antipatterns (Brown, Malveau, McCormick, & Mowbray, 1998), design flaws (Riel, 1996), design characteristics (Whitmire, 1997), or bug patterns (Allen, 2002). Techniques from KDD support the refactoring of software systems (Rech, 2004), and techniques from knowledge management can foster experience-based refactoring (Rech & Ras, 2004).

Quality Defect Discovery

A central research problem in software maintenance is still the inability to change software easily and quickly (Mens & Tourwe, 2004). To improve the quality of their products, organizations often use quality assurance techniques to tackle defects that reduce software quality. The techniques for the discovery of quality defects are based upon several research fields.

- **Software Inspections** (Aurum, Petersson, & Wohlin, 2002; Ciolkowski, Laitenberger, Rombach, Shull, & Perry, 2002), and especially code inspections are concerned with the process of manually inspecting software products in order to find potential ambiguities as well as functional and non-functional problems (Brykczynski, 1999). While the specific evaluation of code fragments is probably more precise than automated techniques, the effort for the inspection is higher, the completeness of an inspection regarding the whole system is smaller, and the number of quality defects searched for is smaller.

- **Software testing** (Liggesmeyer, 2003) and debugging is concerned with the discovery of defects regarding the functionality and reliability as defined in a specification or unit test case in static and dynamic environments.
- **Software product metrics** (Fenton & Neil, 1999) are used in software analysis to measure the complexity, cohesion, coupling, or other characteristics of the software product, which are further analyzed and interpreted to estimate the effort for development or to evaluate the quality of the software product. Tools for software analysis in existence today are used to monitor dynamic or static aspects of software systems in order to manually identify potential problems in the architecture or find sources for negative effects on the quality.

Furthermore, several specific techniques for quality defect discovery already exist (Marinescu, 2004; Rapu, Ducasse, Girba, & Marinescu, 2004). Most of the tools such as Checkstyle, FindBugs, Hammurapi, or PMD analyze the source code of software systems to find violations of project-specific programming guidelines, missing or overcomplicated expressions, as well as potential language-specific functional defects or bug patterns. Nowadays, the Sotograph can identify "architectural smells" that are based on metrics regarding size or coupling (Roock & Lippert, 2005).

But the information from these techniques and the resulting CAPP or *refactoring* activities are typically lost after some time if they are not documented in external documents or *defect management* systems (e.g., bugzilla). And even these external data sources are prone to get lost over several years of maintenance and infrastructure changes. The only information that will not get lost is typically the source code itself.

Refactoring

Beside the development of software systems, the effort for software evolution and maintenance is estimated to amount to 50% to 80% of the overall development cost (Verhoef, 2000). One step in the evolution and development of software systems is the process of reworking parts of the software in order to improve its structure and quality (e.g., maintainability, reliability, usability, etc.), but not its functionality. This process of improving the internal quality of object-oriented software systems in agile software development is called *refactoring* (Fowler, 1999). While refactoring originates in from the agile world, it can, nevertheless, be used in plan-driven (resp. heavyweight) software engineering. In general, refactoring (Fowler, 1999; Mens et al., 2004) is necessary to remove quality defects that are introduced by quick and often unsystematic development.

The primary goal of agile methods is the rapid development of software systems that are continuously adapted to customer requirements without large process overhead. During the last few years, refactoring has become an important part in agile processes for improving the structure of software systems between development cycles. Refactoring is able to reduce the cost, effort, and time-to-market of software systems. Development, maintenance, and reengineering effort are reduced by restructuring existing software systems (on the basis of best practices, design heuristics, and software engineering principles), especially in the process of understanding (the impact of new changes in) a system. A reduction of effort also reduces the length of projects and therefore, cost and time-to-market. Furthermore, refactoring improves product quality and therefore is able to reduce the complexity and size of software systems. Especially in agile software development, methods as well as tools to support refactoring are becoming more and more important (Mens, Demeyer, Du Bois, Stenten, & Van Gorp, 2003).

However, performing manual discovery of quality defects that should be refactored result in either very short or costly refactoring phases. While several automations for refactoring have already been developed (e.g., "extract method" refactoring), the location, analysis, and removal is still an unsystematic, intuitive, and manual process. Today, several techniques and methods exist to support software quality assurance (SQA) on higher levels of abstraction (e.g., requirement inspections) or between development iterations (e.g., testing). Organizations use techniques like refactoring to tackle *quality defects* (i.e., bad smells in code (Beck & Fowler, 1999), architecture smells (Roock et al., 2005), anti-patterns (Brown et al., 1998), design flaws (Riel, 1996; Whitmire, 1997), and software anomalies (IEEE-1044, 1995), etc.) that reduce software quality.

Refactoring does not stop after discovery; even if we had solved the problem of discovering every quality defect possible, the information about the defect, the rationales of whether it is removed (or not), and the refactorings used have to be documented in order to support maintainers and reengineers in later phases. If one knows how to remove a specific quality defect or a group of quality defects, one still needs support, as it is not clear where and under which conditions refactoring activities should be used. Furthermore, product managers need support to organize chains of refactorings and to analyze the impact of changes due to refactorings on the software system. Analogously, quality managers and engineers need information to assess the software quality, identify potential problems, select feasible countermeasures, and plan the refactoring process as well as preventive measures (e.g., code inspections).

Defect Documentation

Today, various repositories exist for documenting of information about defects, incidents, or other issues regarding software changes. This information can be stored in configuration management systems (e.g., CVS, SourceSafe), code reuse repositories (e.g., ReDiscovery, InQuisiX), or *defect management systems*.

The last category is also known as bug tracking (Serrano et al., 2005), issue tracking (Johnson & Dubois, 2003), defect tracking (Fukui, 2002), or source code review systems (Remillard, 2005). They enable a software engineer to record information about the location, causes, effects, or reproducibility of a defect. Typical representatives of defect management systems are open-source variants such as Bugzilla (Serrano et al., 2005), Scarab (Tigris, 2005), Mantis (Mantis, 2005), or TRAC (TRAC, 2005). Commercial versions include Tuppas (Tuppas, 2005), Census from Metaquest (MetaQuest, 2005), JIRA from Atlassian (Atlassian, 2005), or SSM from Force10 (Force10, 2005). These tools are predominantly used in defect handling to describe defects on the lower abstractions of software systems (i.e., source code) (Koru & Tian, 2004) separated from the code.

Defect classification schemes (Freimut, 2001; Pepper, Moreau, & Hennion, 2005) like ODC (Orthogonal Defect Classification) (Chillarege, 1996) are used, for example, in conjunction with these tools to describe the defects and the activity and status a defect is involved in. The ODC process consists of an opening and closing process for defect detection that uses information about the target for further removal activities. Typically, removal activities are executed, but changes, decisions, and experiences are not documented at all—except for small informal comments when the software system is checked into a software repository like CVS.

From our point of view, the direct storage of information about defects, decisions about them, or refactorings applied in the code (as a central point of information) via *annotation languages* such as JavaDoc (Kramer, 1999), doxygen (van Heesch, 2005), or ePyDoc (Loper, 2004) seems to be a more promising solution. The next section describes the relevant background and related work for annotation languages, which are used to record historical information about the evolution of a code fragment (e.g., a method, class, subsystem, etc.).

Source Code Annotation Languages

Annotation languages such as JavaDoc (Kramer, 1999), ePyDoc (ePyDoc, 2005), ProgDOC (Simonis & Weiss, 2003), or Doxygen (van Heesch, 2005) are typically used to describe the characteristics and functionality of code fragments (i.e., classes, methods, packages, etc.) in the source code itself or in additional files. Today several extensions, especially to JavaDoc, are known that enable us to annotate which patterns (Hallum, 2002; Torchiano, 2002), aspects (Sametinger & Riebisch, 2002), or refactorings (Roock & Havenstein, 2002) were or will be used on the source code, and which help us to describe characteristics such as invariants, pre-/ post-conditions, or reviewer names (JSR-260, 2005; Tullmann, 2002). These extensions to the annotation language are called taglets. They are used by doclets in the extraction using, for example, the JavaDoc program. These tools collect the distributed information blocks and generate a (online) documentation rendered in HTML or another file format (e.g., PDF) for better viewing. Typically, these documentations describe the application program interface (API) as a reference for software engineers. Similarly, tools and notations like Xdoclet offer additional tags that are used to generate many artifacts such as XML descriptors or source code. These files are generated from templates using the information provided in the source code and its JavaDoc tags.

Typical content of code annotations is, for example, used to describe the:

- Purpose of a class, field, or method.
- Existence of (functional) defects or workarounds.
- Examples of using the code fragment.

In the following sections and tables, we describe the tags currently available for annotating source code using JavaDoc. JavaDoc is a name for an annotation language as well as the name of a tool from Sun Microsystems to generate API documentation and is currently the industry standard for documenting software systems in Java. The tool uses the tags from the JavaDoc language to generate the API documentation in HTML format. It provides an API for creating doclets and taglets, which allows extending the system with one's own tags (via taglets) and the documentation with additional information (via doclets).

As listed in 0, JavaDoc currently consists of 19 tags that might be used to describe distinguished information (e.g., such as return values of a method) or to format text passages (e.g., to emphasize exemplary source code). The standard tags appear as "`@tag`" and might include inline tags, which appear within curly brackets "`{@ tag}`." Inline tags only appear within, respectively behind, standard tags or in the description field (e.g., "`@pat.name … {@pat.role …}`").

Developers can use the JavaDoc tags when documenting source code in a special comment block by starting it with "`/**`" and ending it with "`*/`." A tag is indicated by using an "`@`" ("at") sign right before the tag name. An example of a JavaDoc comment used for a method is in Box 1.

As an extension to JavaDoc, four refactoring tags were developed in Roock et al. (2002) as described in 0.

Box 1.

/**	Start of JavaDoc comment
* Sorts an array using quicksort	Description of the method
* @author John Doe	Indicate the author
* @param productArray	Describe a parameter
* @return Array The sorted array	Describe the return value
*/	End of JavaDoc comment

Table 1. General tags of the JavaDoc annotation language

Tag	Description	Origin	Type
@author	May appear several times and indicates who has created or modified the code.	JavaDoc 1.0	Context
@param	Describes one parameter of a method (or template class).	JavaDoc 1.0	Function
@return	Describes the returned object of a method.	JavaDoc 1.0	Function
@throws	Describes the (exception-) objects that are thrown by this method.	JavaDoc 1.2	Function
@exception	Synonym for @throws.	JavaDoc 1.0	Function
@version	States the version of this code structure.	JavaDoc 1.0	Context
@since	States the version since when this code was implemented and available to others.	JavaDoc 1.1	Context
@deprecated	Indicates that this code structure should not be used anymore.	JavaDoc 1.0	Status
@see	Adds a comment or link to the "See also" section of the documentation. May link to another part of the documentation (i.e., code).	JavaDoc 1.0	Reference
@serialData	Comments the types and order of data in a serialized form.	JavaDoc 1.2	Context
@serialField	Comments a ObjectStreamField.	JavaDoc 1.2	Context
@serial	Comments default serializable fields.	JavaDoc 1.2	Context
<@code>	Formats text in code font (similar to <code>).	JavaDoc 1.5	Format
<@docRoot>	Represents the relative path to the root of the documentation.	JavaDoc 1.3	Reference
<@inherit-Doc>	Copies the documentation from the nearest inherited code structure.	JavaDoc 1.4	Reference
<@link>	Links to another part of the documentation (i.e., code structure) as the @see tag but stays inline with the text and is formated as "code."	JavaDoc 1.2	Reference
<@linkPlain>	Identical to <@link> but is displayed in normal text format (i.e., not code format).	JavaDoc 1.4	Reference
<@literal>	Displays text without interpreting it as HTML or nested JavaDoc.	JavaDoc 1.5	Format
<@value>	The value of a local static field or of the specified constant in another code.	JavaDoc 1.4	Reference

Table 2. Refactoring tags by Roock et al. (2002)

Tag	Description
@past	Describes the previous version of the signature.
@future	Describes the future signature of the element.
@paramDef	States the default value expected for a parameter. The syntax is @paramDef <parameter> = <value>.
@default	Defines the default implementation of an abstract method.

Table 3. Pattern tags by Torchiano (2002)

Tag	Description
@pat.name	States the standard name of the pattern as defined in (Gamma, Richard, Johnson, & Vlissides, 1994) (and other).
<@pat.role>	Inline-tag of pat.name that describes the part of the pattern that is represented by this element (e.g., "Leaf" in a composite pattern).
@pat.task	Describes the task performed by the pattern or its role.
@pat.use	Describes the use of the pattern or a role, typically by a method.

Table 4. Other tags

Tag	Description
@contract	Defines bounds of a parameter (or other value). Syntax is "@contract <requires> <min> <= <parameter> <= <max>."
@inv, @invariant	States an invariant. Syntax is "@inv <boolean expression>."
@pre	States the precondition for a method.
@post	States the postcondition for a method. This includes information about side effects (e.g., changes to global variables, fields in an object, changes to a parameter, and return values (except if stated in @return).
@issue	Indicates a new requirement or feature that could be implemented. Syntax is @issue [description ...].
@reviewedBy	Indicates a code review for the associated class/interface was completed by a reviewer. Syntax is @reviewedby <name> <date> [notes ...].
@license	Indicates the copyright license used for this code fragment. Syntax is @license [description ...].
@category	Annotates the element with a free attribute / category. @category <category>.
@example	@example <description>.
@tutorial	Link to a tutorial.
@index	Defines the text that should appear in the index created by JavaDoc.
@exclude	States that this element should not be included in the API by the JavaDoc command.
@todo	Indicates that further work has to be done on this element.
@internal	Comments to this element that are internal to the developer or company.
@obsolete	Used if deprecated elements are actually removed from the API.
@threadSafe	Indicates whether this element is threadsafe.
@pattern	Formally describes a pattern existence with the syntax @pattern <pattern name>.<instance name> <role name> <text>.
@aspect	Describes an aspect existence with the syntax @aspect <name> <text>.
@trace	Describes a pattern existence with the syntax @trace <name> <text>.

Table 5. Annotation languages in comparison

Language	Extension	# of Tags	Test Info	Inspection Info	Pattern Info	Refactoring Info
JavaDoc 1.5	Standard	19	No	No	No	No
	Roock et al.	5	Semi-Formal (1)	No	No	Informal (4)
	Torchiano	10	No	No	Semi-Formal (3)	No
	Samet-inger et al.	3	No	No	Informal (3)	No
	Tullmann	4	No	Informal (1)	No	No
	Kramer	3	Informal (3)	No	No	No
	JSR-260	9	No	No	No	No

To note the existence of patterns in a software system as well as the task and role as described in the pattern definitions, several tags were developed by Torchiano (2002) and are listed in 0.

Furthermore, several other groups of annotations exist for various purposes. The following tags are from Roock et al. (2002) (@contract), Kramer (1998) (@inv, @pre, @post), Tullmann (2002) (@issue, @todo, @reviewedBy, @license), Sametinger et al. (2002) (@pattern, @aspect, @trace), and JSR-260 (2005) (@category, @example, @tutorial, @index, @exclude, @todo, @internal, @obsolete, @threadSafe).

The characteristics of source code annotation languages can be differentiated by the number of tags and the formality of their expressiveness. We differentiate between three categories of formality:

1. **Formal:** An explicit and unambiguous specification of the content. A formal tag might include an informal section like a description or note to the formal part (e.g., the tag "param" in JavaDoc has an informal part to describe the meaning of the parameter). In particular, the formal part of a tag must be processable by a computer.
2. **Semi-formal:** A structured or formal representation that is ambiguous or not directly processable by a computer.
3. **Informal:** An unstructured and possibly ambiguous specification of content.

In summary, the tags used in JavaDoc and its extensions can be used to describe characteristics of the source code on a relatively granular or semi-formal level. The processing of these annotations can be used to generate API documentations with additional information about patterns, aspects, or signature changes. The recording of quality defects discovered and refactorings applied as well as rationales or experiences about their application can only be accomplished using free text in the API description.

Furthermore, these annotation languages and their extensions have different target areas in the field of software quality assurance in order to store information about tests, inspections, patterns, and refactorings. 0 shows a comparison of several annotation languages in relevant areas.

Quality Defects and Quality Defect Discovery

The main concern of software quality assurance (SQA) is the efficient and effective development of large, reliable, and high-quality software systems. In agile software development, organizations use techniques like refactoring to tackle "bad smells in code" (Beck et al., 1999), which reduce software qualities such as maintainability, changeability, or reusability. Other groups of defects that do not attack functional, but rather non-functional aspects of a software system are architecture smells (Roock & Lippert, 2004), anti-patterns (Brown et al., 1998), design flaws (Riel, 1996; Whitmire, 1997), and software anomalies in general (IEEE-1044, 1995).

In this chapter, we use the umbrella term *quality defects* (QD) for any defect in software systems that has an effect on software quality (e.g., as defined in ISO 9126), but does not directly affect functionality. Whether the quality defect is automatically discoverable (Dromey, 1996, 2003) or not (Lauesen & Younessi, 1998), an annotation language and method that can be used to support the handling of quality defects should record information about quality characteristics and quality defects in order to represent their status and treatment history. This section will elaborate on this concept and describe several quality defects, their interrelation, symptoms, and effects.

Today, various forms of quality defects exist with different types of granularity. Some target problems in methods and classes, while others describe problems on the architecture or even process levels. In this chapter, we only focus on quality defects on the code level. The representatives on this level are:

- **Code Smells:** The term code smell is an abbreviation of "bad smells in code," which were described in Beck et al. (1999). Today, we have many code smells that are semi-formally described and can be used for manual inspection and discovery. There are at least 38 known code smells with 22 in Fowler (1999), 9 new ones in Wake (2003), 5 new ones in Kerievsky (2005), and 2 in Tourwe and Mens (2003). Code smells are indicators for refactoring and typically include a set of alternative refactoring activities in their description, which might be used to remove them.
- **Architectural Smells:** Very similar to code smells are architectural smells that describe problems on the design level. Yet, the 31 architectural smells described in Roock et al. (2005) do not only apply on the design level but also on the code level. They typically describe problems regarding classes in object-oriented software and interrelations between them.
- **Anti-Patterns:** Design patterns (Gamma et al., 1994) and anti-patterns (Brown et al., 1998) represent typical and reoccurring patterns of good and bad software architectures and were the start of the description of many patterns in diverse software phases and products. While patterns typically state and emphasize a single solution to multiple problems, anti-patterns typically state and emphasize a single problem to multiple solutions. An anti-pattern is a general, proven, and non-beneficial problem (i.e., bad solution) in a software product or process. It strongly classifies the problem that exhibits negative consequences and provides a solution. Built upon similar experiences, these anti-patterns represent "worst-practices" about how to structure or build a software architecture. An example is the "lava flow" anti-pattern, which warns about developing a software system without stopping sometimes and reengineering the system. The larger and older such a software system gets, the more dead code and solidified (bad) decisions it carries along.
- **Bug Patterns:** These patterns are concerned with functional aspects that are typically found in debugging and testing activities. In Allen (2002), 15 bug patterns are described, which describe underlying bugs in a software system.
- **Design Flaws and (Negative) Design Characteristics:** Whitmire (1997) describes nine distinct and measurable characteristics of an object-oriented design. These characteristics

such as "similarity" describe the degree to which two or more classes or domain-level abstractions are similar in terms of their structure, function, behavior, or purpose.

- **Design Heuristics:** Design heuristics provide support on how to construct software systems and, in a way, define quality defects by their absence. They range from the "information hiding" principle to guidelines such as "Eliminate irrelevant classes from your design." There are 61 design heuristics described in Riel (1996) and 14 principles described in Roock et al. (2005).

As listed, there are many quality defects of various granularities and they are described in different forms. To give a more concrete idea of quality, we describe two of them in the following:

1. **Long Method:** In object-oriented programming, one should pay attention to the fact that methods are not too long. The longer a method, the more difficult it is to be understood by a developer. Comprehensibility and readability are negatively affected by the length of a method and thus negatively affect maintainability and testability. Moreover, a short understandable method typically needs less comments than a long one. Another advantage of short methods is the fact that a developer does not constantly scroll and break his reading flow. The most obvious method for discovering long methods is the metric number of lines (LOC) per method. But the question of which method is too long and constitutes a problem is not easily answered. This must either be specified or found by detecting anomalies from the standard distribution. Nevertheless, a method exceeding this threshold value must not necessarily be shortened if other, more important, quality constraints would be negatively affected.
2. **Shotgun Surgery:** This denotes the problem that several classes are always changed in the same group, for example, if the system is adapted to a new database scheme and the same two classes are changed each time. The expandability of the system is thus constrained, and if one class is forgotten during a change, it is more likely to fail. The discovery of shotgun surgery is very difficulty and requires either change metrics or specified rules.

While these problems might not represent problems that have a directly tangible effect on quality, they might become problematic in future evolution or refactoring activities and should be removed as fast as possible—if the time is available. These are only two of many described quality defects. Nevertheless, they show that quality defects describe problems on different levels of

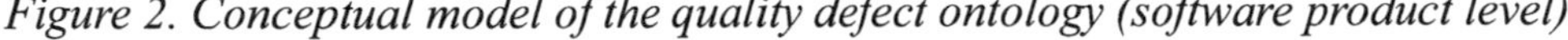

Figure 2. Conceptual model of the quality defect ontology (software product level)

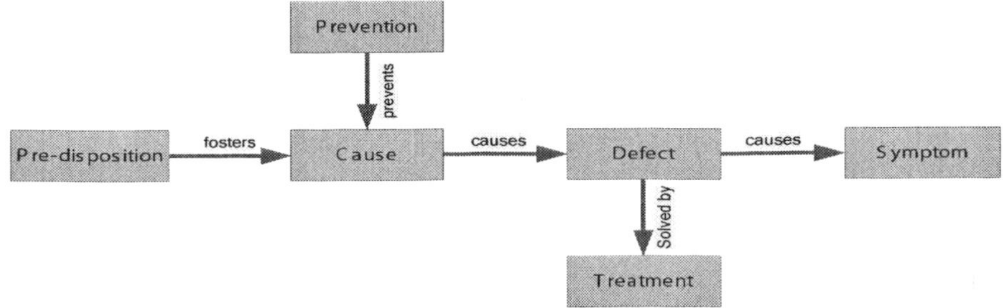

Table 6. Examples for software engineering techniques

	Example 1	Example 2
Predisposition	Data processing system	Lack of good architecture/design
Cause	Large data processing algorithms	Distributed functionality
Defect	"Long method"	"Shotgun surgery"
Side-effects of defect	Increase analyzability effort	Increased maintenance effort
Symptom	Many lines of code	Recurrent changes to the same units
Treatment	Extract method	Inline class(es)
Side-effects of treatment	Increased subroutine calls (worsens performance)	Divergent change
Prevention	Optimize algorithm (phase)	Pattern-based architecture

complexity and might occur in parallel in one situation (i.e., in one code fragment).

0 depicts the general model for the concepts that are used to describe quality defects and that are linked to them. A software system might have *predispositions* that foster or enable the creation of quality defects. These defects themselves have *causes* that are responsible for the defects being integrated into the system. The quality *defects* might have a negative as well as a positive effect on specific qualities and are perceivable via specific *symptoms*. Finally, the defects are solved or removed via specific treatments after they are discovered, or the causes might be prevented by special preventive measures.

In software engineering (SE) and especially in the case of quality defects for a software product, the context of a defect can be described as listed in 0.

In the example on the right side, the predisposition "bad architecture" causes a "cluttered functionality," which results in a "shotgun surgery" defect. This quality defect can be discovered by the symptom of "recurring changes to the same set of software units" and might be removed by the "inline class" refactoring. A prevention of this problem would be a "good" systematic architecture with clear separation of functionality (e.g., in the form of a pattern-based architecture).

Handling of Quality Defects

Typically, during agile development with short iterations, several quality defects are introduced into the software system and are discovered especially in the refactoring phase. To facilitate the annotation of source code and the processing of

Figure 3. The quality defect discovery and handling process model

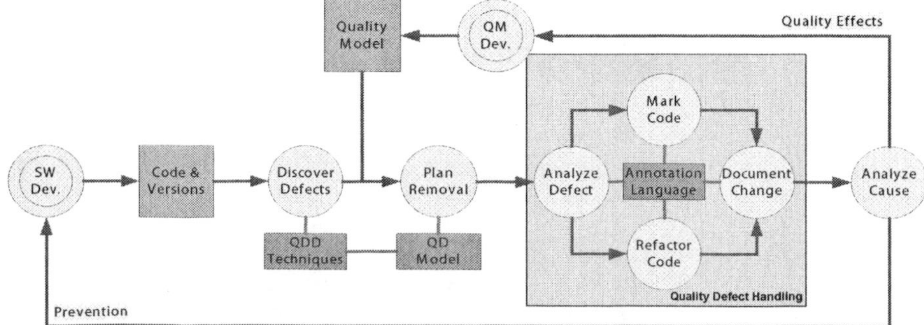

quality defect removal, a mechanism to store the information from the QDD process is required.

The Handling Process

In the following, the general process of quality defect discovery and refactoring is depicted. 0 shows the process model that is either initiated during software development or during a special maintenance (resp. refactoring) activity.

In the execution of the process, the following sub-processes are performed:

- **Discover Defects:** Manual or automatic quality defect discovery techniques are used to analyze the source code and versions thereof from the software repository. Potential quality defects are identified and the affected code (of the most current version) is annotated.
- **Plan Removal:** Based on the discovered quality defects (annotated with a special tag) and a previously defined quality model, a sequential plan for the refactoring of the software system (or part thereof) is constructed.
- **Analyze Defects:** The software engineer processes the list of potential quality defects based on their priority, analyzes the affected software system (or part thereof), and decides if the quality defect is truly present and if the software system can be modified without creating too many new quality defects.
- **Refactor Code:** If the quality defect is to be removed from the software system, the engineer is briefed about the existing quality defects and their rationales as well as about available refactorings, their impact on software quality, and previously made experiences with the kind of quality defect and refactoring at hand.
- **Mark Code:** If a potential quality defect is unavoidable or its removal would have a negative impact on an important quality (e.g., performance), this decision is recorded in the affected part of the software system to prevent future analysis of this part.
- **Document Change:** After the refactoring or marking, the software system is annotated with specific tags about the change or decision, and the experience about the activity is recorded within an experience database (i.e., a database in an experience factory (Basili, Caldiera, & Rombach, 1994b) for storing, formalizing, and generalizing experiences about software development and refactoring activities (e.g., to construct defect patterns from multiple similar defect descriptions)).
- **Analyze Cause:** Statistics, information, and experiences about the existence of quality defects in the software systems are fed back into the early phases of the software development process to prevent or at least reduce their reoccurrence. Preventive measures include, for example, software requirement inspections or goal-oriented training of employees. Furthermore, information about dependencies between qualities, quality defects, and refactorings are fed back into the quality model development process in order to continuously improve the techniques for quality model development.

Decision Support in Handling

In order to support decisions about what to refactor in a software system, we developed several methods and techniques. The following questions act as the guiding theme for the development and enactment of decision-making (i.e., the "plan removal" or "refactor code" phase) as well as understanding (i.e., the "analyze defect" or "document change" phase) in refactoring phases:

- **Decision problem 1:** Which quality defects should be refactored and which might stay in the system?
- **Decision problem 2:** In which sequence should one refactor multiple quality defects in order to minimize effort?
- **Comprehension problem 1:** How does one understand and detect the quality defect in the concrete situation?
- **Comprehension problem 2:** How does one understand the refactoring in the concrete situation and its effect on the software system?
- **Decision problem 3:** Which refactoring should one use if multiple ones are available?
- **Comprehension problem 3:** Which information should one record after refactoring or marking for later evolution, maintenance, or reengineering activities?
- **Decision problem 4:** Did the understanding of the problem or the refactoring result in valuable experience that should be recorded to support later activities (possibly by others)?
- **Comprehension problem 5:** How should one record the experience?

Decision Support in Software Refactoring

Our approach encompasses several methods for supporting the decision of where, when, and in what sequence to refactor a software system as depicted in 0. Beginning from the left upper corner and going counterclockwise, knowledge about quality defects from defect discovery processes is used to retrieve experiences associated with similar defects from previous refactorings. These experiences are used to handle quality defects in the defect removal phase. Additionally, suitable experiences are augmented by so-called micro-didactical arrangements (MDA) (Ras, Avram, Waterson, & Weibelzahl, 2005), which initiate learning processes and aim at improving the understandability, applicability, and adaptability of the experience in the specific context.

As shown in 0, we define six phases, based on the quality improvement paradigm (QIP) (Basili, Caldiera, & Rombach, 1994a), for the continuous handling of quality defects. In contrast to the quality defect handling process as depicted in 0, these phases are not concerned with quality defects in a specific product, but with the learning process about the quality defects themselves and their

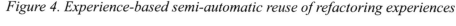

Figure 4. Experience-based semi-automatic reuse of refactoring experiences

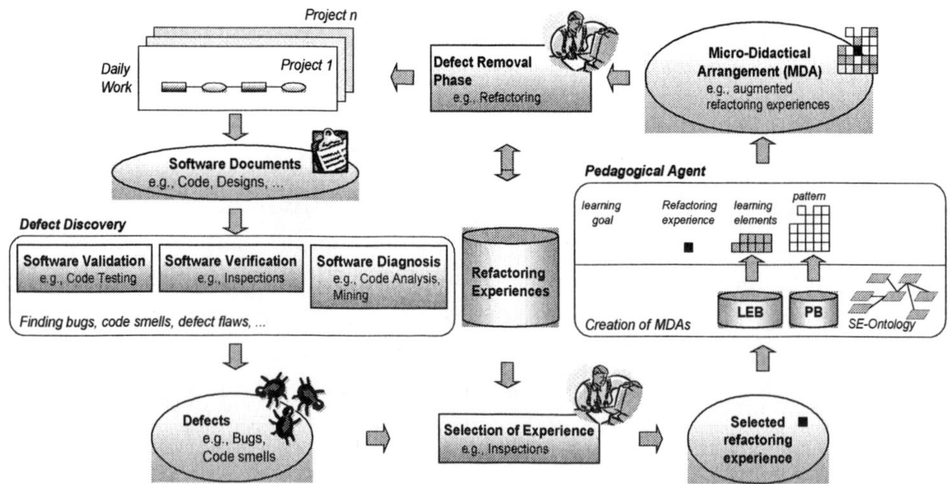

effect on the software qualities. 0 represents the realizations of phase 2 ("discover defect"), phase 3 ("plan removal"), and phase 4 (the "quality defect handling" block).

In 0, we first start with the definition of the quality model consisting of qualities that should be monitored and improved. For example, this may result in different goals (i.e., quality aspects), as reusability demands more flexibility or "openness," while maintainability requires more simplicity. Phase 2 is concerned with the measurement and preprocessing of the source code to build a basis for quality defect discovery (i.e., "discover defects"). Results from the discovery process (i.e., quality defects) are represented and prioritized to plan the refactoring in phase 3 (i.e., "plan removal"). Here, the responsible person has to decide which refactorings have to be executed (i.e., "analyze defect") in what configuration and sequence, in order to minimize work (e.g., change conflicts) and maximize the effect on a specific quality. In phase 4, the refactoring itself is (or is not) applied to the software system (i.e., "Refactor Code" or "Mark Code") by the developer, which results in an improved product. Phase 5 compares the improved product with the original product to detect changes and their impact on the remaining system (i.e., "analyze cause"). Finally, in phase 6, we document the experiences and data about the refactoring activity, changes to the software system, and other effects in order to learn from

Figure 5. Quality-driven refactoring

our work and continuously improve the model of relationships between quality, refactorings, and quality defects.

As indicated previously, the KDD sub-processes are grouped in phase 2. We select source code from a specific build, preprocess the code, and store the results in the code warehouse, analyze the data to discover quality defects, discover deviations from average behavior, cluster code blocks with severe or multiple quality defects, and represent discovered and prioritized quality defects to the user.

An Example of DS for QDD

For example, we may detect a method in an object-oriented software system that has a length of 300 LOC. As described in Fowler (1999), this is a code smell called long method. A long method is a problem especially in maintenance phases, as the responsible maintainer will have a hard time understanding the function of this method.

One suitable refactoring for the mentioned code smell might be the refactoring simply called extract method: the source code of the long method is reviewed to detect blocks that can be encapsulated into new (sub-)methods. Experiences with the extract method refactoring are used to support the decision on where, when, how, and if the refactoring has to been implemented. For example, the developer might remark that every block of code that has a common meaning, and could be commented respectively, could also be extracted into several smaller methods. Furthermore, the developer might note that the extraction of (sub-) methods, from methods implementing complex algorithms, can affect the performance requirements of the software system and therefore might not be applicable.

Additionally, the generation of new methods might create another smell called "large class" (i.e., the presence of too many methods in a class), which might complicate the case even further. Finally, the new experiences are annotated by the developer and stored in the refactoring experience base.

While this example only touches a simple quality defect and refactoring, more complex refactorings influence inheritance relations or introduce design patterns (Fowler, 1999).

An Annotation Language to Support Quality Defect Handling

This section describes, defines, and explains a language that will be used to annotate code fragments that are either contaminated by quality defects or checked by a software engineer and cleared of quality defects. As described in the background section, several annotation languages for the annotation of source code already exist that are not adequate. This new language is used to keep decisions about quality defects persistent and over time builds a "medical history" of the source code fragment (e.g., a class).

Goals and Characteristics of Annotation Languages

All annotation languages represent a basis for describing additional information about the software system directly at the code level. *Target groups* (or users) for the documentation/annotation language are:

- **Developers,** who want to use the source code and acquire information via the API descriptions (e.g., for software libraries).
- **Testers,** who want to develop test cases and need information about the pre- and post-conditions as well as the functionality to be tested.
- **Maintainers,** who want to evolve the system and need information about existing quality defects, rationales for their persistence (e.g., refactoring would cause loss of performance), or past refactorings (e.g., to update the software documentation such as design documents).

In our case, an annotation language that is targeted at supporting the handling of quality defects should encompass several key aspects. The requirements for such an annotation language should cover uses such as:

- **Annotate change** for later understanding by the same and other readers (e.g., maintainers).
- **Mark fragment** that a quality defect is detected but can or must stay in the system.
- **Note membership** in a larger quality defect or refactoring activity that encompassed multiple code fragments for later impact analyses.
- **Annotate quality aspects** for later reuse, etc.
- **Tag additional information** in the code fragment freely or based on a classification (e.g., "problematic class," "quicksort algorithm," "part of subsystem X") to support later reuse or maintenance/reengineering activities (similar to social software or Web 2.0 approaches).

We identified the following information blocks of an annotation language that should be recorded with an annotation language and that are based on the six knowledge types from knowledge management (Mason, 2005):

- **Know-what:** Record the currently present *quality defects* that were found manually or automatically.
- **Know-how:** Record the *transformation history* (similar to the medical history of a human patient).
- **Know-why:** Record the *rationales* why a refactoring was applied or why a quality defect is still present in order to prevent

recurrent defect analysis or refactoring attempts.

- **Know-where:** Record the *location* in the annotated code as well as associated code fragments that where changed as well.
- **Know-who:** Record the tool or *person* (i.e., developer or maintainer) who applied the refactoring.
- **Know-when:** Record the time or *version* when the quality defect was found or the refactoring was applied. This could also be used to define a trigger when a refactoring has to be applied (e.g., if several other (larger) refactorings or design decision have to be made).
- **Context:** Record the frame of reference or context in which the quality defect was discovered. This includes especially the quality model used to decide which quality defect has a higher priority over other quality defects.

The following requirements for tags and other constructs in such an annotation language to support refactoring and maintenance activities are:

- **Unambiguous:** The names of tags, quality defects, refactorings, or other reoccurring terms should be unique and used consistently throughout the system.
- **Machine-readable:** The syntax of tags should be formal, exact, and consistent to avoid confusion and enable the interpretation and usage by supporting systems (e.g., defect discovery tools).
- **Local completeness:** The power of the syntax should be large enough to cover all existing cases. Full comprehensiveness is probably not possible except by allowing informal free text attributes.
- **Flexibility:** The syntax should not limit the extension by new tags or tag attributes.
- **Independence:** Tags should describe information that is mutually exclusive, and the occurrence of two or more tags should be independent from one another.

Beside the additional documentation of the software system, the annotation language will increase the semantic coupling between code fragments and reduce the presence of quality defects such as "shotgun surgery."

RAL: The Refactoring Annotation Language

The refactoring annotation language (RAL) is used to record the currently existing quality characteristics, symptoms, defects, and refactoring of a code fragment regarding a specific quality model. Furthermore, it is used to store the rationales and treatment history (e.g., sequence of refactorings).

In the following tables, the core set of tags from RAL are described based on the JavaDoc syntax and using existing JavaDoc and supportive tags, that are used in the description and will be described after the core tags. Information blocks starting with a double cross "#" indicate an ID or standardized term from an external, controlled vocabulary or taxonomy.

A symptom tag as defined in 0 describes a metric or characteristic of the code fragment and is used as an indicator for the statistical or rule-based identification of quality defects. The tag acts as an annotation of a specific symptom from a controlled vocabulary in order to have a unique identifier and a reference for further information about the symptom. The since tag from JavaDoc is used to identify the version based on which the quality symptom was first calculated.

The quality defect as defined in 0 represents a code smell, antipattern, etc. present in this code fragment. It is used to annotate a specific quality defect from a controlled vocabulary in order to have a unique identifier and reference for more information about a specific quality defect type and potential treatments. The since tag from Ja-

Table 7. The @symptom tag

Tag Syntax	@symptom <#Symptom-ID> <@value value> <@since #version>
Example	@symptom "LOC" @value "732" @since 1.2

Table 8. The @defect tag

Tag Syntax	@defect <#QD-ID> <@since #version> <@status #Status> <@rationale text>
Example	@defect "Long Method" @since 1.2 @status "untreated"

Table 9. The @refactoring tag

Tag Syntax	@refactoring <#Refactoring-ID> <@rationale text> <@status #Status> <@link fragment> <@ author name>
Example	@refactoring "Extract Method" "Applied as quality model rates maintainability higher than performance" @status "treated" @link "ExtractedMethod" @author "John Doe"

Table 10. The @quality-model tag

Tag Syntax:	@quality-model Name <@see file>
Example	@quality-model "QM-Dep1-Project2" @see <A href=http://www.sw-quality.de/QM-Dep1-Project2.pdf>

Table 11. Support tags

Tag	Description
@status	"@status #status" indicates the current status of the superior tag or source code using the vocabulary "discovered," "inWork," "treated," or (deliberately) "untreated."
@rationale	"@rationale text" declares a rationale about the existence or status of the superior tag or source code.

vaDoc is used to identify the version where the quality defect was first noticed.

A refactoring tag as defined in 0 is a description of a single refactoring that was applied for removing one or more quality defects. Optionally, a project-internal URI to other code fragments directly affected by the refactoring (e.g., if two classes interchange a method during the same refactoring) can be stated.

The quality model tag as defined in 0 is used as a reference to the quality model that defines which quality characteristics are important, what priority or decision model lies beneath, and which quality defects are relevant to a specific part of the software system. Optionally, it refers to a URI of a file containing the specific (machine-readable) quality model.

The supportive tags used in the previous tag descriptions are given in 0.

Depending on the processor that would render a quality documentation from these tags, some tags might be used only once and inherited by lower levels. For example, the quality model tag needs only be stated once (e.g., for the whole project) or twice (e.g., for the client and server part) in a software system.

RAL is primarily used to annotate source code. Therefore, in order to annotate documents of higher abstraction, like UML-based design documents (e.g., platform-independent models in MDA) using the XMI Format or formal requirement documents, similar languages (probably based on other languages such as JavaDoc) need to be defined.

Handling Quality Defects Using RAL

Software annotation languages like JavaDoc or Doxygen and extensions like RAL can now be used to document the functionality, structure, quality, and treatment history of the software system at the code level. The formal basis of the language enables tools to read and write this information automatically to generate special documents or trigger specific actions.

The core tags `@symptom`, `@defect`, and `@refactoring` build on top of each other and might be recorded by several different tools. This enables the intertwined cooperation of different tools, each with a specific focus, such as to calculate metrics or to discover quality defects. For example, one tool might measure the source code and its versions to extract numerical and historical information and write it into `@symptom` tags (e.g., lines of code). Another tool might analyze this information to infer quality defects (e.g., "long method") that are recorded in `@defect` tags. Finally, a tool might offer refactorings to a developer or a maintainer during his work and note applied refactorings or rationales in explicit `@refactoring` tags.

Developers and maintainers of a software system are supported in the handling of quality defects in the following activities:

- Repetitive refactoring of a specific kind of quality defect (e.g., "large method"), as they do not have to switch between different defects or refactoring concepts.
- Reuse of knowledge about the refactoring of specific quality defects to judge new quality defects.
- Recapitulation of the change history of the code fragment to update software documentation such as design documents.
- Retrieval of information about persons who developed or refactored this part of the system and should know about its purpose and functionality.
- Product or quality managers of the software system might use the information to:
- Evaluate the quality based on information extracted via the tags about the amount or distribution of quality defects.
- Analyze specific dates or groups of persons that might have introduced specific kinds of quality defects and might need further training.

SUMMARY AND OUTLOOK

Agile software development methods were invented to minimize the risk of developing low-quality software systems with rigid process-based methods. They impose as little overhead as possible in order to develop software as fast as possible and with continuous feedback from the customers. To assure quality, agile software development organizations use activities such as refactoring between development iterations. *Refactoring,* or the restructuring of a software system without changing its behavior, is necessary to remove *quality defects* (i.e., bad smells in code, architecture smells, anti-patterns, design flaws, software anomalies, etc.) that are introduced by quick and often unsystematic development. However, the effort for the manual discovery of these quality defects results in either incomplete or costly refactoring phases. Furthermore, software quality assurance methods seem to ignore their recurring application.

In this chapter, we described a process for the recurring and sustainable discovery, handling, and treatment of quality defects in software systems. We described the complexity of the discovery and handling of quality defects in object-oriented source code to support the software refactoring process. Based on the formal definition of quality defects, we gave examples of how to support the recurring and sustainable handling of quality defects. The annotation language presented is used to store information about quality defects found in source code and represents the defect and treatment history of a part of a software system. The process and annotation language can not only be used to support quality defect discovery processes, but also has the potential to be applied in testing and inspection processes.

Recapitulating, we specified an annotation language that can be used in agile software maintenance and refactoring to record information about quality defects, refactorings, and rationales about them. Similar annotation languages such as JavaDoc or doxygen as well as third party extensions are not able to encode this information in a machine-readable and unambiguous format.

The proposed framework including the handling process promises systematic and semi-automatic support of refactoring activities for developers, maintainers, and quality managers. The approach for recording quality defects and code transformations in order to monitor refactoring activities will make maintenance activities simpler and increase overall software quality. Likewise, the user monitors daily builds of the software to detect code smells, identical quality defects, or groups thereof, and initiates repetitive refactoring activities, minimizing effort caused by task switches.

REQUIREMENTS FOR QUALITY DEFECT HANDLING IN AGILE SE

When building systems and languages for quality defect handling in agile software development, several requirements should be kept in mind.

The annotation language in the form of a code annotation language like JavaDoc or in the form of an external documentation such as a Defect Tracking system or a Wiki should be integrated into the programming language used and into the development environment. If it is not integrated, the information might easily be lost due to the high workload and time constraints in agile development. Especially in an agile environment, the developers, testers, and maintainers should be burdened with as little additional effort as possible.

Therefore, the more formal the descriptions of an annotation language are and the more information can be extracted from the code and development environment (e.g., from the refactoring techniques), the less information is required from the developers.

OUTLOOK

The trend in research is to increase automation of the mentioned processes in order to support the developers with automated refactoring or defect discovery systems.

We expect to further assist software engineers and managers in their work and in decision making. One current research task is the development of taglets and doclets to generate specific evolution documents. Furthermore, we are working on the analysis and synthesis of discovery techniques with statistical and analytical methods based on textual, structural, numerical, and historical information.

Although we can record and use this information in several applications, we currently do not know if the amount of information might overwhelm or annoy the developer and maintainer. If dozens of quality defects are found and additional refactorings are recorded, this might be confusing and should be hidden (e.g., in an editor of the IDE) from the developer. Very old information (e.g., from previous releases of the software) might even be removed and stored in an external document or database.

REFERENCES

Allen, E. (2002). *Bug patterns in Java*. New York; Berkeley, CA: Apress.

Atlassian. (2005). *JIRA Web site*. Retrieved October 6, 2005, from http://www.atlassian.com/software/jira/

Aurum, A., Petersson, H., & Wohlin, C. (2002). State-of-the-art: Software inspections after 25 years. Software testing. *Verification and Reliability, 12*(3), 133-154.

Basili, V. R., Caldiera, G., & Rombach, D. (1994a). The goal question metric approach. In *Encyclopedia of software engineering* (1st ed., pp. 528-532). New York: John Wiley & Son.

Basili, V. R., Caldiera, G., & Rombach, H. D. (1994b). *Experience factory*. In J. J. Marciniak (Ed.), *Encyclopedia of software engineering* (Vol. 1, pp. 469-476). New York: John Wiley & Sons.

Beck, K. (1999). *eXtreme programming eXplained: Embrace change*. Reading, MA: Addison-Wesley.

Beck, K., & Fowler, M. (1999). *Bad smells in code*. In G. Booch, I. Jacobson, & J. Rumbaugh (Eds.), Refactoring: Improving the design of existing code (1st ed., pp. 75-88). Addison-Wesley Object Technology Series.

Brown, W. J., Malveau, R. C., McCormick, H. W., & Mowbray, T. J. (1998). *AntiPatterns: Refactoring software, architectures, and projects in crisis*. New York: John Wiley & Sons, Inc.

Brykczynski, B. (1999). A survey of software inspection checklists. *Software Engineering Notes, 24*(1), 82-89.

Chillarege, R. (1996). Orthogonal defect classification. In M. R. Lyu (Ed.), *Handbook of software reliability engineering* (pp. xxv, 850 p.). New York: IEEE Computer Society Press.

Ciolkowski, M., Laitenberger, O., Rombach, D., Shull, F., & Perry, D. (2002). *Software inspections, reviews, and walkthroughs*. Paper presented at the 24th International Conference on Software Engineering (ICSE 2002), New York, USA, Soc.

Dromey, R. G. (1996). Cornering the chimera. *IEEE Software, 13*(1), 33-43.

Dromey, R. G. (2003). Software quality—Prevention versus cure? *Software Quality Journal, 11*(3), 197-210.

ePyDoc. (2005). *Epydoc Web site*. Retrieved May 10, 2005, from http://epydoc.sourceforge.net/

Fenton, N. E., & Neil, M. (1999). Software metrics: Successes, failures, and new directions. *Journal of Systems and Software, 47*(2-3), 149-157.

Force10. (2005). *Software support management system (SSM) Web site*. Retrieved October 6, 2005, from http://www.f10software.com/

Fowler, M. (1999). *Refactoring: Improving the design of existing code* (1st ed.). Addison-Wesley.

Freimut, B. (2001). *Developing and using defect classification schemes* (Technical Report No. IESE-Report No. 072.01/E). Kaiserslautern: Fraunhofer IESE.

Fukui, S. (2002). *Introduction of the software configuration management team and defect tracking system for global distributed development*. Paper presented at the 7th European Conference on Software Quality (ECSQ 2002), Helsinki, Finland, June 9-13, 2002.

Gamma, E., Richard, H., Johnson, R., & Vlissides, J. (1994). *Design patterns: Elements of reusable object-oriented software* (3rd ed., Vol. 5). Addison-Wesley.

Hallum, A. M. (2002). *Documenting patterns*. Unpublished Master Thesis, Norges Teknisk-Naturvitenskapelige Universitet.

IEEE-1044. (1995). IEEE guide to classification for software anomalies. IEEE Std 1044.1-1995.

Johnson, J. N., & Dubois, P. F. (2003). Issue tracking. *Computing in Science & Engineering, 5*(6), 717, November-December.

JSR-260. (2005). *Javadoc Tag Technology Update (JSR-260)*. Retrieved October 6, 2005, from http://www.jcp.org/en/jsr/detail?id=260

Kerievsky, J. (2005). *Refactoring to patterns*. Boston: Addison-Wesley.

Koru, A. G., & Tian, J. (2004). Defect handling in medium and large open source projects. *IEEE Software*, *21*(4), 54-61.

Kramer, D. (1999, September 12-14). *API documentation from source code comments: A case study of Javadoc*. Paper presented at the 17th International Conference on Computer Documentation (SIGDOC 99), New Orleans, LA.

Kramer, R. (1998). iContract—The Java(tm) Design by Contract(tm) Tool. In *Technology of object-oriented languages and systems, TOOLS 26* (pp. 295-307). Santa Barbara, CA: IEEE Computer Society.

Lauesen, S., & Younessi, H. (1998). Is software quality visible in the code? *IEEE Software*, *15*(4), 69-73.

Liggesmeyer, P. (2003). Testing safety-critical software in theory and practice: A summary. *IT Information Technology*, *45*(1), 39-45.

Loper, E. (2004). *Epydoc: API documentation extraction in python*. Retrieved from http://epydoc.sourceforge.net/pycon-epydoc.pdf

Mantis. (2005). *Mantis Web site*. Retrieved October 6, 2005, from http://www.mantisbt.org/

Marinescu, R. (2004, September 11-14). *Detection strategies: Metrics-based rules for detecting design flaws*. Paper presented at the 20th International Conference on Software Maintenance, Chicago, IL.

Mason, J. (2005). From e-learning to e-knowledge. In M. Rao (Ed.), *Knowledge management tools and techniques* (pp. 320-328). London: Elsevier.

Mens, T., Demeyer, S., Du Bois, B., Stenten, H., & Van Gorp, P. (2003). Refactoring: Current research and future trends. *Electronic Notes in Theoretical Computer Science*, *82*(3), 17.

Mens, T., & Tourwe, T. (2004). A survey of software refactoring. *IEEE Transactions on Software Engineering*, *30*(2), 126-139.

MetaQuest. (2005). *Census Web site*. Retrieved October 6, 2005, from http://www.metaquest.com/Solutions/BugTracking/BugTracking.html

Pepper, D., Moreau, O., & Hennion, G. (2005, April 11-12). *Inline automated defect classification: A novel approach to defect management*. Paper presented at the IEEE/SEMI Advanced Semiconductor Manufacturing Conference and Workshop, Munich, Germany.

Rapu, D., Ducasse, S., Girba, T., & Marinescu, R. (2004). *Using history information to improve design flaws detection*. Paper presented at the 8th European Conference on Software Maintenance and Reengineering, Tampere, Finland.

Ras, E., Avram, G., Waterson, P., & Weibelzahl, S. (2005). Using Weblogs for knowledge sharing and learning in information spaces. *Journal of Universal Computer Science*, *11*(3), 394-409.

Rech, J. (2004). *Towards knowledge discovery in software repositories to support refactoring*. Paper presented at the Workshop on Knowledge Oriented Maintenance (KOM) at SEKE 2004, Banff, Canada.

Rech, J., & Ras, E. (2004). *Experience-based refactoring for goal-oriented software quality improvement*. Paper presented at the 1st International Workshop on Software Quality (SOQUA 2004), Erfurt, Germany.

Remillard, J. (2005). Source code review systems. *IEEE Software*, *22*(1), 74-77.

Riel, A. J. (1996). *Object-oriented design heuristics*. Reading, MA: Addison-Wesley.

Roock, S., & Havenstein, A. (2002). *Refactoring tags for automatic refactoring of framework dependent applications.* Paper presented at the Extreme Programming Conference XP 2002, Villasimius, Cagliari, Italy.

Roock, S., & Lippert, M. (2004). *Refactorings in großen Softwareprojekten: Komplexe Restrukturierungen erfolgreich durchführen (in German).* Heidelberg: dpunkt Verlag.

Roock, S., & Lippert, M. (2005). *Refactoring in large software projects.* John Wiley & Sons.

Sametinger, J., & Riebisch, M. (2002, March 11-13). *Evolution support by homogeneously documenting patterns, aspects, and traces.* Paper presented at the 6th European Conference on Software Maintenance and Reengineering, Budapest, Hungary.

Serrano, N., & Ciordia, I. (2005). Bugzilla, ITracker, and other bug trackers. *IEEE Software, 22*(2), 11-13.

Simonis, V., & Weiss, R. (2003, July 9-12). *PROGDOC—a new program documentation system.* Paper presented at the 5th International Andrei Ershov Memorial Conference (PSI 2003) Perspectives of System Informatics, Novosibirsk, Russia.

Tigris. (2005). *Scarab Web site.* Retrieved October 6, 2005, from http://scarab.tigris.org/

Torchiano, M. (2002, October 3-6). *Documenting pattern use in Java programs.* Paper presented at the Proceedings of the International Conference on Software Maintenance (ICSM), Montreal, Que., Canada.

Tourwe, T., & Mens, T. (2003). Identifying refactoring opportunities using logic meta programming. IEEE Computer, Reengineering Forum; Univ. Sannio. In *Proceedings 7th European Conference on Software Maintenance and Reengineering,* Los Alamitos, CA (xi+2420 2091-2100). IEEE Comput. Soc.

TRAC. (2005). *TRAC Web site.* Retrieved October 6, 2005, from http://projects.edgewall.com/trac/

Tullmann, P. (2002). *Pat's taglet collection.* Retrieved October 6, 2005, from http://www.tullmann.org/pat/taglets/

Tuppas. (2005). *Tuppas Web site.* Retrieved October 6, 2005, from http://www.tuppas.com/Defects.htm

van Heesch, D. (2005). *Doxygen—a documentation system.* Retrieved from http://www.doxygen.org/

Verhoef, C. (2000, September 5-7). *How to implement the future?* Paper presented at the Proceedings of the 26th EUROMICRO Conference (EUROMICRO2000), Maastricht, The Netherlands.

Wake, W. C. (2003). *Refactoring workbook* (1st ed.). Pearson Education.

Whitmire, S. A. (1997). *Object-oriented design measurement.* New York: John Wiley & Sons.

Chapter VI
Agile Quality Assurance Techniques for GUI-Based Applications

Atif Memon
University of Maryland, USA

Qing Xie
Accenture Technology Labs, USA

ABSTRACT

This chapter motivates the need for new agile model-based testing mechanisms that can keep pace with agile software development/evolution. A new concentric loop-based technique, which effectively utilizes resources during iterative development, is presented. The tightest loop is called crash testing, which operates on each code check-in of the software. The second loop is called smoke testing, which operates on each day's build. The third and outermost loop is called the "comprehensive testing" loop, which is executed after a major version of the software is available. Because rapid testing of software with a graphical-user interface (GUI) front-end is extremely complex, and GUI development lends itself well to agile processes, the GUI domain is used throughout the chapter as an example. The GUI model used to realize the concentric-loop technique is described in detail.

INTRODUCTION

Agile software development has had a significant impact on the development of software applications that contain a graphical user interface (GUI). GUIs are by far the most popular means used to interact with today's software. A GUI uses one or more metaphors for objects familiar in real life such as buttons, menus, a desktop, the view through a window, a trash can, and the physical layout in a room. Objects of a GUI include elements such as windows, pull-down menus, buttons, scroll bars, iconic images, and wizards. A software user performs *events* to interact with the GUI, manipulating GUI objects as one would real objects. For example, dragging an item, discarding an object by dropping it in a trash can, and selecting items from a menu are all familiar events available in today's GUI. These events may cause changes to the state of the software that may be reflected by a change in the appearance of one or more GUI objects.

Recognizing the importance of GUIs, software developers are dedicating an increasingly large portion of software code to implementing GUIs--up to 60% of the total software code (Mahajan & Shneiderman, 1996; Myers, 1993a, 1995a, 1995b; Myers & Olsen, 1994). The widespread use of GUIs is leading to the construction of increasingly complex GUIs. Their use in safety-critical systems is also growing (Wick, Shehad, & Hajare, 1993).

GUI-based applications lend themselves to the core practices of agile development, namely simple planning, short iteration, and frequent customer feedback. GUI developers work closely with customers iteratively enhancing the GUI via feedback. Although agile processes apply perfectly to GUI software, integration testing of the GUI for overall functional correctness remains complex, resource intensive, and *ad hoc*. Consequently, GUI software remains largely untested during the iterative development cycle. Adequately testing a GUI is required to help ensure the safety, robustness, and usability of an entire software system (Myers, Hollan, & Cruz, 1996). Testing is, in general, labor and resource intensive, accounting for 50-60% of the total cost of software development (Gray, 2003; Perry, 1995). GUI testing is especially difficult today because GUIs have characteristics different from those of traditional software, and thus, techniques typically applied to software testing are not adequate.

Testing the correctness of a GUI is difficult for a number of reasons. First of all, the space of possible interactions with a GUI is enormous, in that each sequence of GUI events can result in a different state, and each GUI event may need to be evaluated in all of these states (Memon, Pollack, & Soffa, 1999, 2000b). The large number of possible states results in a large number of input permutations (White, 1996) requiring extensive testing. A related problem is to determine the coverage of a set of test cases (Memon, Soffa, & Pollack, 2001c). For conventional software, coverage is measured using the amount and type of underlying code exercised. These measures do not work well for GUI testing because what matters is not only how much of the code is tested, but in how many different possible states of the software each piece of code is tested. An important aspect of GUI testing is verification of its state at each step of test case execution (Memon, Pollack, & Soffa, 2000a). An incorrect GUI state can lead to an unexpected screen, making further execution of the test case useless since events in the test case may not match the corresponding GUI elements on the screen. Thus, the execution of the test case must be terminated as soon as an error is detected. Also, if verification checks are not inserted at each step, it may become difficult to identify the actual cause of the error. Finally, regression testing presents special challenges for GUIs because the input-output mapping does not remain constant across successive versions of the software (Memon & Soffa, 2003e; Myers, Olsen, & Bonar, 1993b). Regression testing is especially important for GUIs since GUI development typically uses an agile model (Kaddah, 1993; Kaster, 1991; Mulligan, Altom, & Simkin, 1991; Nielsen, 1993).

The most common way to test a GUI is to wait until the iterative development has ended and the GUI has "stabilized." Testers then use capture/replay tools (Hicinbothom & Zachary, 1993) such as WinRunner (http://mercuryinteractive.com) (Memon, 2003a) to test the new *major GUI version release*. A tester uses these tools in two phases: a capture and then a replay phase. During the capture phase, a tester manually interacts with the GUI being tested, performing events. The tool records the interactions; the tester also visually "asserts" that a part of the GUI's response/state be stored with the test case as "expected output" (Memon & Xie, 2004c; Memon, Banerjee, & Nagarajan, 2003d). The recorded test cases are replayed automatically on (a modified version of) the GUI using the replay part of the tool. The "assertions" are used to check whether the GUI executed correctly. Another way to test a GUI is by *programming* the test cases (and expected output) using tools (Finsterwalder, 2001; White, 2001) such as extensions of *JUnit* including *JFCUnit*, *Abbot*, *Pounder*, and *Jemmy Module* (http://junit.org/news/extension/gui/index.htm). The previous techniques require a significant amount of manual effort, typically yielding a small number of test cases. The result is an inadequately tested GUI (Memon, Nagarajan, & Xie, 2005a). Moreover, during iterative development, developers waste time fixing bugs that they encounter in later development cycles; these bugs could have been detected earlier if the GUI had been tested iteratively.

The agile nature of GUI development requires the development of new GUI testing techniques that are themselves agile in that they quickly test each increment of the GUI during development. This chapter presents a process with supporting tools for continuous integration testing of GUI-based applications; this process connects modern model-based GUI testing techniques with the needs of agile software development. The key idea of this process is to create concentric testing loops, each with specific GUI testing goals, requirements, and resource usage. Instances of three such loops are presented. The tightest loop called the *crash testing* loop operates on each code check-in (e.g., using CVS) of the GUI software (Xie & Memon, 2005). It is executed very frequently and hence is designed to be very inexpensive. The goal is to perform a quick-and-dirty, fully automatic integration test of the GUI software. Software crashes are reported back to the developer within minutes of the check-in. The second loop is called the *smoke testing* loop, which operates on each day's GUI build (Memon et al., 2005a; Memon & Xie, 2004b; Memon & Xie, 2005b; Memon, Banerjee, Hashish, & Nagarajan, 2003b). It is executed nightly/daily and hence is designed to complete within 8-10 hours. The goal of this loop is to do functional "reference testing" of the newly integrated version of the GUI. Differences between the outputs of the previous (yesterday's) build and the new build are reported to developers. The third, and outermost

Figure 1. Different loops of continuous GUI integration testing

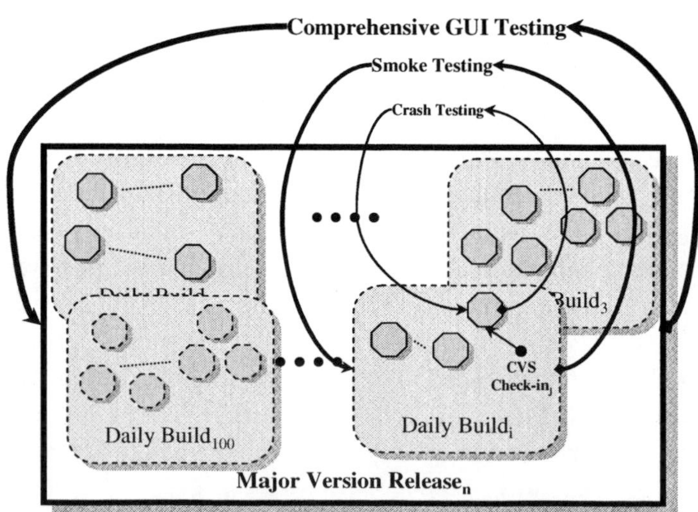

loop is called the "comprehensive GUI testing" loop. It is executed after a major version of the GUI is available. The goal of this loop is to conduct comprehensive GUI integration testing, and hence is the most expensive. Major problems in the GUI software are reported. An overview of this process is shown in Figure 1. The small octagons represent frequent CVS code check-ins. The encompassing rectangles with rounded corners represent daily increments of the GUI. The large rectangle represents the major GUI version. The three loops discussed earlier are shown operating on these software artifacts.

A novel feature of the continuous testing process is a GUI model that is obtained by using automated techniques that employ reverse engineering (Memon, Banerjee, & Nagarajan, 2003c). This model is then used to generate test cases, create descriptions of expected execution behavior, and evaluate the adequacy of the generated test cases. Automated test executors "play" these test cases on the GUI and report errors.

The specific contributions of this work include:

1. Three distinctive product- and stakeholder-oriented, novelty (agile) approaches, and techniques that may be applied to the broad class of event-driven software applications.
2. Comprehensive theoretical and practical coverage of testing in the context of agile quality.

The remainder of this chapter will present an overview of existing approaches used for GUI testing and describe the continuous testing process, including the three concentric testing loops, the GUI model used for automated testing, and future trends.

BACKGROUND

Software Testing

We now give an overview of software testing techniques. The goal of testing is to detect the presence of errors in programs by executing the programs on well-chosen input data. An error is said to be present when either (1) the program's output is not consistent with the specifications, or (2) the test designer determines that the specifications are incorrect. Detection of errors may lead to changes in the software or its specifications. These changes then create the need for re-testing.

Testing requires that test cases be executed on the software under test (SUT) and the software's output be compared with the expected output by using a test oracle. The input and the expected output are a part of the test suite. The test suite is composed of tests each of which is a triple <*identifier, input, output*>, where *identifier* identifies the test, *input* is the input for that execution of the program, and *output* is the expected output for this input (Rothermel & Harrold, 1997). The entire testing process for software systems is done using test suites.

Information about the software is needed to generate the test suite. This information may be available in the form of formal specifications or derived from the software's structure leading to the following classification of testing.

- Black-box testing (also called *functional testing* (Beizer, 1990) or *testing to specifications*): A technique that does not consider the actual software code when generating test cases. The software is treated as a black-box. It is subjected to inputs and the output is verified for conformance to specified behavior. Test generators that support black-box testing require that the software specifications be given as rules and procedures. Examples of black-box test techniques are equivalence class partitioning, boundary value analysis, and cause-effect graphing.

- White-box testing (also called *glass-box testing* (Beizer, 1990) or *testing to code*): A technique that considers the actual implementation code for test case generation. For example, a *path oriented* test case generator selects a program's execution path and generates input data for executing the program along that path. Other popular techniques make use of the program's branch structure, program statements, code slices, and control flow graphs (CFG).

No single technique is sufficient for complete testing of a software system. Any practical testing solution must use a combination of techniques to check different aspects of the program.

Gui Testing Steps

Although GUIs have characteristics such as user events for input and graphical output that are different from those of conventional software and thus require the development of different testing techniques, the overall process of testing GUIs is similar to that of testing conventional software. The testing steps for conventional software, extended for GUIs, follow:

- **Determine What to Test:** During this first step of testing, coverage criteria, which are sets of rules used to determine what to test in a software application, are employed. In GUIs, a coverage criterion may require that each event be executed to determine whether it behaves correctly.
- **Generate Test Input:** The test input is an important part of the test case and is constructed from the software's specifications and/or from the structure of the software. For GUIs, the test input consists of events such as mouse clicks, menu selections, and object manipulation actions.
- **Generate Expected Output:** Test oracles generate the expected output, which is used to determine whether or not the software executed correctly during testing. A test oracle is a mechanism that determines whether or not the output from the software is equivalent to the expected output. In GUIs, the expected output includes screen snapshots and positions and titles of windows.
- **Execute Test Cases and Verify Output:** Test cases are executed on the software and its output is compared with the expected output. Execution of the GUI's test case is done by performing all the input events specified in the test case and comparing the GUI's output to the expected output as given by the test oracles.
- **Determine if the GUI was Adequately Tested:** Once all the test cases have been executed on the implemented software, the software is analyzed to check which of its parts were actually tested. In GUIs, such an analysis is needed to identify the events and the resulting GUI states that were tested and those that were missed. Note that this step is important because it may not always be possible to test in a GUI implementation what is required by the coverage criteria.

After testing, problems are identified in the software and corrected. Modifications then lead to regression testing (i.e., re-testing of the changed software).

- **Perform Regression Testing:** Regression testing is used to help ensure the correctness of the modified parts of the software as well as to establish confidence that changes have not adversely affected previously tested parts. A regression test suite is developed that consists of (1) a subset of the original test cases to retest parts of the original software that may have been affected by modifications, and (2) new test cases to test affected parts of the software, not tested by the selected test cases. In GUIs, regression testing involves analyzing the changes to the layout of GUI objects, selecting test cases that should be rerun, as well as generating new test cases.

Any GUI testing method must perform all of the previous steps. As mentioned earlier, GUI test designers typically rely on capture/replay tools to test GUIs (Hicinbothom et al., 1993). The process involved in using these tools is largely manual making GUI testing slow and expensive.

A few research efforts have addressed the automation of test case generation for GUIs. A finite state machine (FSM) based modeling approach is proposed by Clarke (1998). However, FSM models have been found to have scaling problems when applied to GUI test case generation. Slight variations such as variable finite state machine (*VFSM*) models have been proposed by Shehady and Siewiorek (1997). These techniques help scalability but require that verification checks be inserted manually at points determined by the test designer.

Test cases have been generated to mimic *novice users* (Kasik & George, 1996). The approach uses an expert to generate the initial path manually and then use genetic algorithm techniques to generate longer paths. The assumption is that experts take a more direct path when solving a problem using GUIs whereas novice users often take longer paths. Although useful for generating multiple scripts, the technique relies on an expert to generate the initial script. The final test suite depends largely on the paths taken by the expert user. The idea is using a task and generating an initial script may be better handled by using planning, since multiple scripts may be generated automatically according to some predetermined coverage criteria.

Agile Testing

There are several feedback-based mechanisms to help manage the quality of software applications developed using agile techniques. These mechanisms improve the quality of the software via continuous, rapid QA during iterative improvement. They differ in the level of detail of feedback that they provide to targeted developers, their thoroughness, their frequency of execution, and their speed of execution. For example, some mechanisms (e.g., integrated with CVS) provide immediate feedback at change-commit time by running select test cases, which form the commit validation suite. Developers can immediately see the consequences of their changes. For example, developers of NetBeans perform several quick (Web-enabled) validation steps when checking into the NetBeans CVS repository (http://www.netbeans.org/community/guidelines/commit.html). In fact, some Web-based systems such as Aegis (http://aegis.sourceforge.net/) will not allow a developer to commit changes unless all commit-validation tests have passed. This mechanism ensures that changes will not stop the software from "working" when they are integrated into the software baseline. Other, slower mechanisms include "daily building and smoke testing" that execute more thorough test cases on a regular (e.g., nightly) basis at central server sites. Developers do not get instant feedback; rather they are e-mailed the results of the nightly builds and smoke tests. Another, still higher level of continuous QA support is provided by mechanisms such as Skoll (Memon et al., 2004a) that continuously run test cases, for days and even weeks on several builds (stable and beta) of the evolving software using user-contributed resources. All these mechanisms are useful, in that they detect defects early during iterative development. Moreover, since the feedback is directed towards specific developers (e.g., those who made the latest modifications), QA is implicitly and efficiently distributed.

THE AGILE GUI TESTING PROCESS

Recent research in automated GUI testing has focused on developing techniques that use GUI models for testing. This section consolidates these techniques to provide an end-to-end solution that addresses the challenges of agile GUI development. This section presents details of the overall process shown in Figure 1 and its loops.

Users interact with a GUI by performing *events* on some widgets such as clicking on a button, opening a menu, and dragging an icon. During GUI testing, test cases consisting of sequences of events are executed on the GUI. Earlier work demonstrated that simply executing each event in isolation is not enough for effective GUI testing (Memon et al., 2001c). *Test oracles* are used to determine whether the test cases *failed* or *passed* (Memon et al., 2000a). The agile concentric loops differ in the way they generate test cases and test oracles. These differences lead to varying degrees of effort required by the test designer during testing. Each of these loops, their goals, and requirements are discussed in subsequent sections.

Crash Testing

The goal of crash testing is to create test cases that can quickly test major parts of the GUI fully automatically without any human intervention. More specifically, crash testing generates and executes test cases and oracles that satisfy the following requirements.

- The crash test cases should be generated quickly on the fly and executed. The test cases are not saved as a suite; rather, a throwaway set of test cases that require no maintenance is obtained.
- The test cases should broadly cover the GUI's entire functionality.
- It is expected that new changes will be made to the GUI before the crash testing process is complete. Hence, the crash testing process will be terminated and restarted each time a new change is checked-in. The crash test cases should detect major problems in a short time interval.
- The test oracle simply needs to determine whether the software *crashed* (i.e., terminated unexpectedly during test case execution).

Details of the crash testing process have been presented in earlier reported work (Xie et al., 2005). An empirical study presented therein showed that the crash testing process is efficient in that it can be performed fully automatically, and useful, in that it helped to detect major GUI integration problems. The feedback from crash testing is quickly provided to the specific developer who checked in the latest GUI changes. The developer can debug the code and resubmit the changes before the problems effect other developers' productivity.

Smoke Testing

Smoke testing is more complex than crash testing and hence requires additional effort on the part of the test designer. It also executes for a longer period of time. Moreover, the smoke testing process is not simply looking for crashes—rather its goal is to determine whether the software "broke" during its latest modifications. More specifically, GUI smoke testing has to produce test cases that satisfy the following requirements:

- The smoke test cases should be generated and executed quickly (i.e., in one night).
- The test cases should provide adequate coverage of the GUI's functionality. As is the case with smoke test cases of conventional software, the goal is to raise a "something is wrong here" alarm by checking that GUI

events and event-interactions execute correctly.

- As the GUI is modified, many of the test cases should remain usable. Earlier work showed that GUI test cases are very sensitive to GUI changes (Memon, Pollack, & Soffa, 2001a). The goal here is to design test cases that are robust, in that a majority of them remain unaffected by changes to the GUI.
- The smoke test suite should be divisible into parts that can be run (in parallel) on different machines.
- The test oracle should compare the current version's output with that of the previous version and report differences.

Feasibility studies involving smoke testing (Memon et al., 2005a; Memon et al., 2005b) showed that GUI smoke testing is effective at detecting a large number of faults. Testers have to examine the test results and manually eliminate false positives, which may arise due to changes made to the GUI. The combination of smoke and crash testing ensures that "crash bugs" will not be transmitted to the smoke testing loop. Such bugs usually lead to a large number of failed and unexecuted test cases, causing substantial delays.

Comprehensive GUI Testing

Comprehensive GUI testing is the most expensive, and hence the least frequent executed testing loop during GUI evolution. Since GUI development is iterative, valuable resources are conserved by employing a model-based approach for this loop. Hence, this loop must produce test cases that satisfy the following requirements:

- The test cases should cover the entire functionality of the GUI.
- The test cases should be generated from a model of the GUI. As the GUI evolves, this model is updated by the developers.
- The test oracle should also be generated from the same model. Hence, the model should encode the expected GUI behavior.
- During test execution, the test cases should be capable of detecting all deviations from the GUI specifications represented in the model.

Earlier work used a specialized encoding of GUI events (in terms of preconditions and effects) to generate test cases (Memon et al., 2001a) and test oracles (Memon et al., 2000a). An AI Planner was used to generate the test cases. The test cases revealed all deviations from the specifications.

GUI Model

As previously noted, all the test cases and oracles for the agile GUI testing process are generated automatically using model-based techniques. This section describes the model and how it is used for test automation.

The representation of the model (called the *event-flow model*) contains two parts. The first part encodes each event in terms of *preconditions* (i.e., the state in which the event may be executed), and *effects* (i.e., the changes to the state after the event has executed). The second part represents all possible sequences of events that can be executed on the GUI as a set of directed graphs. Both these parts play important roles for various GUI testing tasks. The preconditions/effects are used for goal-directed test-case generation (Memon et al., 2001a) and test oracle creation (Memon et al., 2000a) for comprehensive testing. The second part is used for graph-traversal based test-case generation (Memon et al., 2005b) for smoke and crash testing. The test oracle for smoke and crash testing does not need to be derived from the GUI model. In case of smoke testing, the oracle looks for differences between the previous and modified GUIs. In case of crash testing, the oracle is hand-coded to look for software crashes.

Modeling Events

An important part of the event-flow model is the behavior of each event in terms of how it modifies the *state* of a GUI when it is executed. Intuitively, the state of the GUI is the collective state of each of its widgets (e.g., buttons, menus) and containers (e.g., frames, windows) that contain other widgets (these widgets and containers will be called GUI objects). Each GUI object is represented by a set of *properties* of those objects (background color, font, caption, etc.).

Formally, a GUI is modeled at a particular time t in terms of:

- Its **objects** $O = \{o_1, o_2, ..., o_m\}$, and
- The **properties** $P = \{p_1, p_2, ..., p_m\}$ of those objects. Each property p_i is represented as a binary Boolean relation, where the name of the relation is the property name, the first argument is the object $o_1 \in O$ and the second argument is the value of the property. Figure 2(a) shows the structure of properties. The property value is a constant drawn from a set associated with the property in question: for instance, the property "background-color" has an associated set of values, {white, yellow, pink, etc.}. Figure 2(b) shows a button object called Button1. One of its properties is called Caption and its current value is "Cancel."

Figure 2(a). The structure of properties, and (b) A Button *object with associated properties*

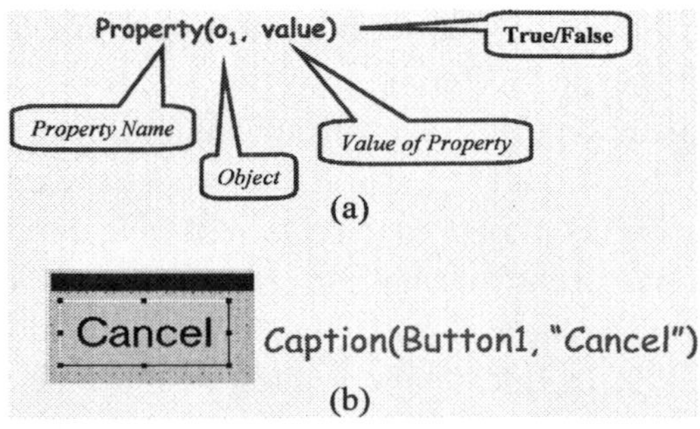

The set of objects and their properties can be used to create a model of the *state* of the GUI.

Definition: The *state* of a GUI at a particular time t is the set P of all the properties of all the objects O that the GUI contains.

A *complete* description of the state would contain information about the types of *all* the objects currently extant in the GUI, as well as *all* of the properties of each of those objects. For example, consider the Open GUI shown in Figure 3(a). This GUI contains several objects, three of which are explicitly labeled; for each, a small subset of its properties is shown. The state of the GUI, partially shown in Figure 3(b), contains all the properties of all the objects in Open.

Events performed on the GUI change its state. Events are modeled as state transducers.

Definition: The *events* $E = \{e_1, e_2, ..., e_n\}$ associated with a GUI are functions from one state of the GUI to another state of the GUI.

Since events may be performed on different types of objects, in different contexts, yielding different behavior, they are parameterized with objects and property values. For example, an event set-background-color(w, x) may be defined in terms of a window w and color x. The parameters w and x may take specific values in the context of a particular GUI execution. As shown in Figure 4, whenever the event set-background-color(w19, yellow) is executed in a state in which window w19 is open, the background color of w19 should become yellow (or stay yellow if it already was), and no other properties of the GUI should change. This example illustrates that, typically, events can only be executed in some states; set-background-color(w19, yellow) cannot be executed when window w19 is not open.

Figure 3(a). The Open *GUI with three objects explicitly labeled and their associated properties, and (b) the State of the* Open *GUI*

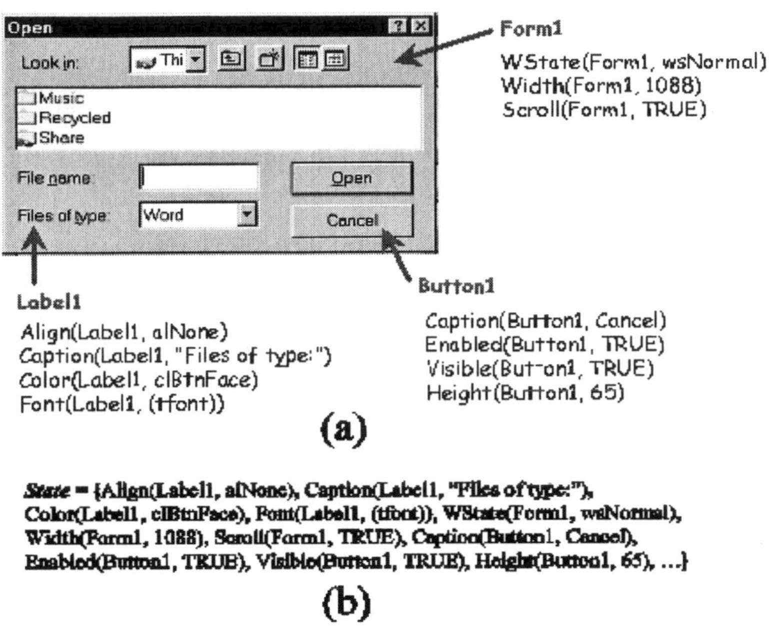

Figure 4. An Event Changes the State of the GUI.

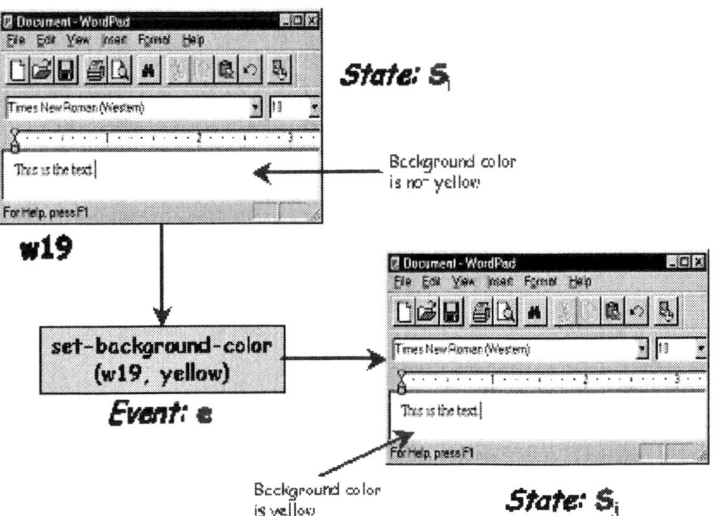

It is of course infeasible to give exhaustive specifications of the state mapping for each event. In principle, as there is no limit to the number of objects a GUI can contain at any point in time, there can be infinitely many states of the GUI. Of course in practice, there are memory limits on the machine on which the GUI is running, and hence only finitely many states are actually possible, but the number of possible states will be extremely large. Hence, GUI events are represented using *operators*, which specify their preconditions and effects:

Definition: An *operator* is a triple <Name, Preconditions, Effects> where:

- Name identifies an event and its parameters.
- Preconditions is a set of positive ground literals $p(arg_1, arg_2)$, where p is a property (*i.e.*, $p \in P$). *Pre*(*Op*) represents the set of preconditions for operator *Op*. A literal is a sentence without conjunction, disjunction or implication; a literal is ground when all of its arguments are bound; and a positive literal is one that is not negated. An operator is applicable in any state S_i in which all the literals in *Pre*(*Op*) are true.
- Effects is also a set of positive or negative ground literals $p(arg_1, arg_2)$, where p is a property (*i.e.*, $p \in P$). *Eff*(*Op*) represents the set of effects for operator *Op*. In the resulting state S_j, all of the positive literals in *Eff*(*Op*) will be true, as will all the literals that were true in S_i except for those that appear as negative literals in *Eff*(*Op*).

For example, the following operator represents the set-background-color event discussed earlier:

- **Name:** set-background-color(wX: window, Col: Color)
- **Preconditions:** current(wX,TRUE), background-color(wX, oldCol), oldCol ≠ Col
- **Effects:** background-color(wX, Col) where current and background-color are properties of window objects.

The previous representation for encoding operators is the same as what is standardly used in the AI planning literature (Pednault, 1989; Weld, 1994; Weld, 1999). This representation has been adopted for GUI testing because of its power to express complex actions.

Generating Test Cases for the Comprehensive Testing Loop

Test case generation for the comprehensive testing loop leverages previous work on using AI planning (Memon et al., 2001a). Because of this previous work, the operators are described in the PDDL language that is used by AI planners. Planning is a goal-directed search technique used to find sequences of actions to accomplish a specific task. For the purpose of test-case generation, given a task (encoded as a pair of initial and goal states) and a set of actions (encoded as a set of operators), the planner returns a sequence of instantiated actions that, when executed, will transform the initial state to the goal state. This sequence is the test case. If no such sequence exists then the operators cannot be used for the task and thus the planner returns "no plan."

Creating Test Oracle for the Comprehensive Testing Loop

The comprehensive GUI testing loop contains the most complex test oracle. A test oracle is a mechanism that determines whether a piece of software executed correctly for a test case. The test oracle may either be automated or manual; in both cases, the actual output is compared to a presumably correct expected output. Earlier work (Memon et al., 2000a) presented the design for a GUI test oracle; it contains three parts (1) an *execution monitor* that extracts the actual state of a GUI using reverse engineering technology (Memon et al., 2003c) as a test case is executed on it, (2) an *oracle procedure* that uses *set equality* to compare the actual state with *oracle information* (i.e., the expected state), (3) the oracle information for a test case $<S_0, e_1; e_2; ...; e_n>$ is defined as a sequence of states $S_1; S_2; ...; S_n$ such that S_i is the expected state of the GUI after event e_i is executed.

Operators are used to obtain the oracle information. Recall that the event-flow model represents events as state transducers. The preconditions-effects-style of encoding the operators makes it fairly straightforward to derive the expected state. Given the GUI in state S_{i-1}, the next state S_i (i.e., the expected state after event e_i is executed) may be computed using the effects of the operator Op representing event e_i via simple additions and deletions to the list of properties representing state S_{i-1}. The next state is obtained from the current state S_{i-1} and ***Eff*(*Op*)** as follows:

1. Delete all literals in S_{i-1} that unify with a negated literal in *Eff*(*Op*), and
2. Add all positive literals in *Eff*(*Op*).

Going back to the example of the GUI in Figure 4 in which the following properties are true before the event is performed: background-color(w19, blue), current(w19,TRUE). Application of the previous operator, with variables bound as set-background-color(w19, yellow), would lead to the following state: background-color(w19, yellow), current(w19,TRUE) (i.e., the background color of window w19 would change from blue to yellow). During test-case execution, this expected state is used to check the correctness of the GUI's actual state.

Note that a literal that is not explicitly added or deleted in the operator's effects remains unchanged (i.e., it persists in the resulting state). This persistence assumption built into the method for computing the result state is called the "STRIPS assumption." A complete formal semantics for operators making the STRIPS assumption has been developed by Lifschitz (1986). It turns out that this persistence assumption makes the operators compact and easy to code since there is no need to consider unchanged widgets and their properties.

Thus, given a test case for the comprehensive testing loop and the operators of the GUI, the expected state can be derived by iterative application of the two previous steps. This expected state is used to create a test oracle during test-case execution.

Modeling Event Interactions

The goal is to represent all possible event interactions in the GUI. Such a representation of the event interaction space is used for automated test case generation for the smoke and crash testing loops. Intuitively, a graph model of the GUI is constructed. Each vertex represents an event

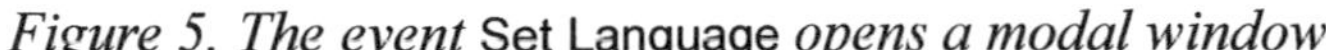

Figure 5. The event Set Language *opens a modal window*

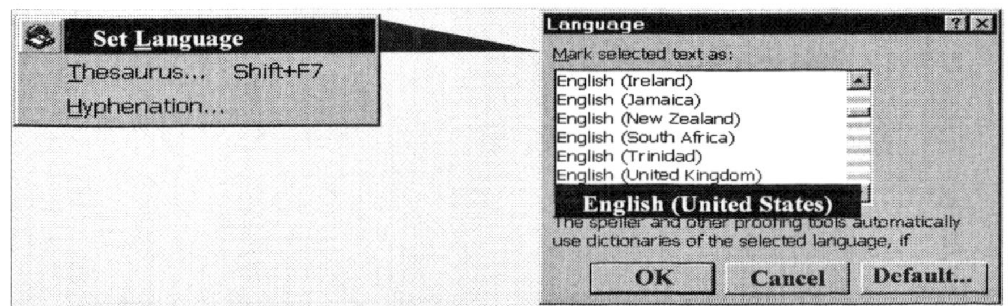

Figure 6. The event Replace *opens a modeless window*

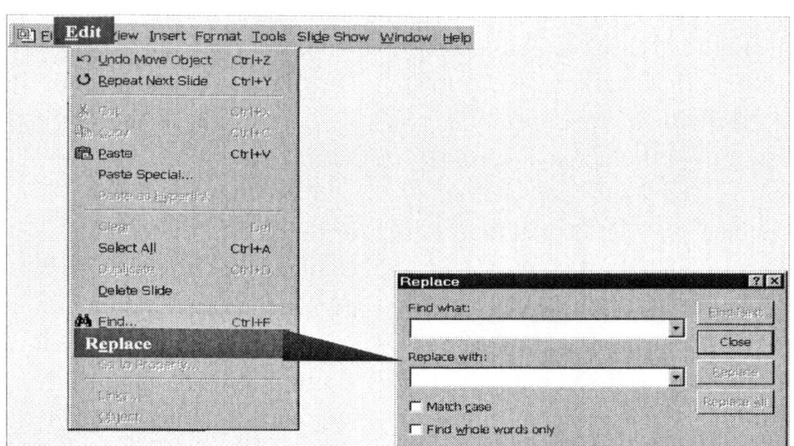

(e.g., click-on-Edit, click-on-Paste). In subsequent discussion for brevity, the names of events are abbreviated (e.g., Edit and Paste). An edge from vertex *x* to vertex *y* shows that an event *y* can be performed immediately after event *x*. This graph is analogous to a control-flow graph in which vertices represent program statements (in some cases basic blocks) and edges represent possible execution ordering between the statements. A state machine model that is equivalent to this graph can be constructed—the state would capture the possible events that can be executed on the GUI at any instant; transitions cause state changes whenever the number and type of available events changes. For a pathological GUI that has no restrictions on event ordering and no windows/menus, such a graph would be fully connected. In practice, however, GUIs are hierarchical, and this hierarchy may be exploited to identify groups of GUI events that may be modeled in isolation. One hierarchy of the GUI and the one used in this research is obtained by examining the structure of *modal windows* (Standard GUI terminology; see details at msdn.microsoft.com/library/en-us/vbcon/html/vbtskdisplayingmodelessform.asp and documents.wolfram.com/v4/AddOns/JLink/1.2.7.3.html.) in the GUI.

Definition: A *modal window* is a GUI window that, once invoked, monopolizes the GUI interaction, restricting the focus of the user to a specific range of events within the window, until the window is explicitly terminated.

The language selection window is an example of a modal window in MS Word. As Figure 5 shows, when the user performs the event Set Language, a window entitled Language opens and the user spends time selecting the language, and finally explicitly terminates the interaction by either performing OK or Cancel.

Other windows in the GUI that do not restrict the user's focus are called *modeless windows*; they merely expand the set of GUI events available to the user. For example, in the MS Word software, performing the event Replace opens a modeless window entitled Replace (Figure 6).

At all times during interaction with the GUI, the user interacts with events within a *modal dialog*. This modal dialog consists of a modal window *X* and a set of modeless windows that have been invoked, either directly or indirectly from *X*. The modal dialog remains in place until *X* is explicitly terminated.

Definition: A *modal dialog (MD)* is an ordered pair (RF, UF) where RF represents a modal window in terms of its events and UF is a set whose elements represent modeless windows also in terms of their events. Each element of UF is invoked (i.e., window is opened) either by an event in RF or UF.

Note that, by definition, a GUI user cannot interleave events of one modal dialog with events of other modal dialogs; the user must either explicitly terminate the currently active modal dialog or invoke another modal dialog to execute events in different dialogs. This property of modal dialogs enables the decomposition of a GUI into parts—each part can be tested separately. As will be seen later, interactions between these parts are modeled separately (as an integration tree) so that the GUI can be tested for these interactions.

Event interactions within a modal dialog may be represented as an event-flow graph. Intuitively, an *event-flow graph* of a modal dialog represents all possible event sequences that can be executed in the dialog. Formally, an event-flow graph is defined as follows.

Definition: An *event-flow graph* for a modal dialog *MD* is a 4-tuple $<V, E, B, I>$ where:

1. V is a set of vertices representing all the events in *MD*. Each $v \in V$ represents an event in *MD*.
2. $E \subseteq V \times V$ is a set of directed edges between vertices. Event e_j follows e_i iff e_j may be performed immediately after e_i. An edge $(v_x, v_y) \in E$ iff the event represented by v_y follows the event represented by v_x.
3. $B \subseteq V$ is a set of vertices representing those events of *MD* that are available to the user when the modal dialog is first invoked.
4. $B \subseteq V$ is the set of events that open other modal dialogs.

An example of an event-flow graph for the Main modal dialog (i.e., the modal dialog that is available to the user when the application is launched) of MS WordPad is shown in Figure 7. To increase readability of the event-flow graph, some edges have been omitted. Instead, labeled circles have been used as connectors to sets of events. The legend shows the set of events represented by each circle. For example, an edge from Save to ① is a compact represent of a collection of edges from the event Save to each element in the set represented by ①. At the top of the figure are the vertices, File, Edit, View, Insert, Format, and Help, that represent the pull-down menu of MS WordPad. They are events that are available when the Main modal dialog is first invoked; they form the set **B**. Once File has been performed in WordPad any of the events in ① may be performed; there are edges in the event-flow graph from File to each of these events. Note that Open is shown as a dashed oval. This notation is used for events that open other modal dialogs; About and Contents are also similar events. Hence, for this modal dialog **I** = {*all events shown with dashed ovals*}. Other events such as Save, Cut, Copy, and Paste are all events that don't open windows; they interact with the underlying software.

Once all the modal dialogs of the GUI have been represented as event-flow graphs, the remaining step is to identify interactions between modal dialogs. A structure called an *integration tree* is constructed to identify interactions (invocations) between modal dialogs.

Definition: Modal dialog MD_x invokes modal dialog MD_y if MD_x contains an event e_x that invokes MD_y.

Intuitively, the integration tree shows the invokes relationship among all the modal dialogs in a GUI. Formally, an integration tree is defined as:

Definition: An integration tree is a triple < N, R, B >, where N is the set of modal dialogs in the GUI and R is a designated modal dialog called the Main modal dialog. B is the set of directed edges showing the invokes relation between modal dia-

Figure 7. Event-flow graph for the Main *modal dialog of MS WordPad*

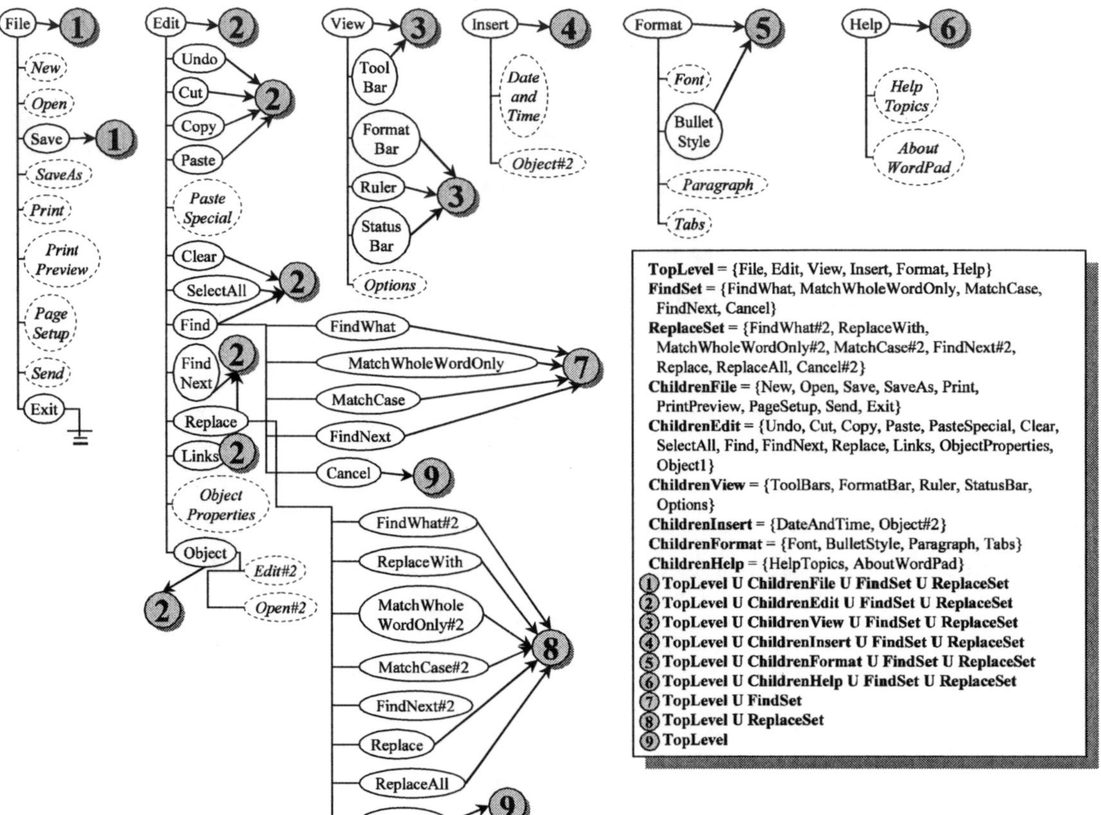

logs (i.e., $(MD_x, MD_y) \in$ B iff MD_x invokes MD_y, where MD_x and MD_y are both modal dialogs).

Figure 8 shows an example of an integration tree representing a part of the MS WordPad's GUI. The vertices represent the modal dialogs of the MS WordPad GUI and the edges represent the invokes relationship between the modal dialogs. The tree in Figure 8 has an edge from Main to FileOpen showing that Main contains an event, namely Open (see Figure 7) that invokes FileOpen.

This decomposition of the GUI makes the overall testing process intuitive for the test designer since the test designer can focus on a specific part of the GUI. Moreover, it simplifies the design of the algorithms and makes the overall testing process more efficient.

Developing the event-flow model manually can be tedious and error-prone. Therefore, a tool called the "GUI Ripper" has been developed to automatically obtain event-flow graphs and the integration tree. A detailed discussion of the tool is beyond the scope of this chapter; the interested reader is referred to previously published work (Memon et al., 2003c) for details. In short, the GUI Ripper combines reverse engineering techniques with the algorithms presented in previous sections to automatically construct the event-flow graphs and integration tree. During "GUI Ripping," the GUI application is executed automatically; the application's windows are opened in a depth-first manner. The GUI Ripper extracts all the widgets and their properties from the GUI. During the

Figure 8. An integration tree for a part of MS WordPad

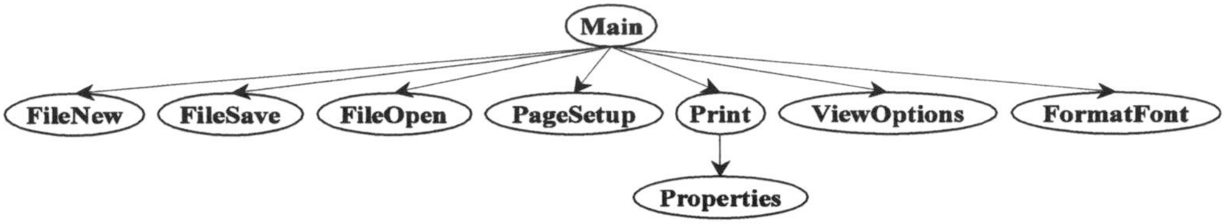

reverse engineering process, in addition to widget properties, additional key attributes of each widget are recovered (e.g., whether it is enabled, it opens a modal/modeless window, it opens a menu, it closes a window, it is a button, it is an editable text-field). These attributes are used to construct the event-flow graphs and integration tree.

As can be imagined, the GUI Ripper is not perfect (i.e., parts of the retrieved information may be incomplete/incorrect). Common examples include (1) missing windows in certain cases (e.g., if the button that opens that window is disabled during GUI Ripping), (2) failure to recognize that a button closes a window, and (3) incorrectly identifying a modal window as a modeless window or vise versa. The specific problems that are encountered depend on the platform used to implement the GUI. For example, for GUIs implemented using Java Swing, the ripper is unable to retrieve the contents of the "Print" dialog; in MS Windows, is unable to correctly identify modal/modeless windows. Recognizing that such problems may occur during reverse engineering, tools have been developed to manually "edit" the event-flow graphs and integration tree and fix these problems.

A test designer also does not have to code each operator from scratch since the reverse engineering technique creates operator templates and fills-in those preconditions and effects that describe the structure of the GUI. Such preconditions and effects are automatically derived during the reverse engineering process in a matter of seconds. Note that there are no errors in these templates since the structure has already been manually examined and corrected in the event-flow graphs and integration trees. The number of operators is the same as the number of events in the GUI, since there is exactly one operator per executable event.

Obtaining Test Cases for Smoke and Crash Testing Loops

Test case generation for the smoke and crash testing loops employ graph traversal of the event-flow graph. More specifically, a GUI test case is of the form $<S_0, e_1; e_2; ...; e_n>$, where S_0 is a state of the GUI in which the event sequence $e_1; e_2; ...; e_n$ is executed. The simplest way to generate test cases is for a tester to start from one of the vertices in the set B of the main modal dialog's event-flow graph. Note that these events are available to a GUI user as soon as the GUI is invoked. The event corresponding to the chosen vertex becomes the first event in the test case. The tester can then use one of the outgoing edges from this vertex to perform an adjacent event. The tester can continue this process generating many test cases. Note that a tester who uses a capture/replay tool to generate test cases is actually executing this process manually without using our formal models.

As noted earlier, if performed manually (using capture/replay tools), the previous process is extremely labor intensive. With the event-flow model, numerous graph-traversal techniques may be used to automate it. The order in which the events are covered will yield different types of test cases. For smoke and crash testing, the tester must generate test cases that (1) cover all the events in the GUI at least once, and (2) cover all pairs of event-interactions at least once. In terms of event-flow graphs, all the edges should be covered by the test cases, thereby ensuring that all events and event interactions are covered. Similar types of such techniques have been used in previous work (Memon et al., 2004b). As mentioned earlier, the GUI model is not needed for test oracle creation for the smoke and crash testing loops.

All the previous techniques have been put together to realize the agile GUI testing process shown in Figure 1.

FUTURE TRENDS

Although this chapter has presented the agile testing concept in terms of GUIs, there is a clear need to extend this work to other event-driven applications, which are becoming increasingly popular; testing these applications faces many of the challenges mentioned earlier for GUIs. Numerous researchers have already started to model various classes of software using their event-driven nature. For example, Welsh, Culler, and Brewer (2001) have modeled Web applications as a network of event-driven stages connected by queues; Duarte, Martins, Domingos, and Preguia (2003) have described an extensible network based event dissemination framework; Gu and Nahrstedt (2001) have presented an event-driven middleware framework; Cheong, Liebman, Liu, and Zhao (2003) have presented a model for event-driven embedded systems; Sliwa (2003) has described how event-driven architectures may be used to develop complex component-based systems; Holzmann and Smith (1999) have modeled device drivers as event-driven systems; and Carloganu and Raguideau (2002) have described an event-driven simulation system. Researchers need to extend the ideas presented in this chapter to the general class of event-driven software.

Similarly, although this work has been presented using event-flow graphs, it is applicable to software that can be modeled using state-machine models. Indeed a state machine model that is equivalent to an event-flow graph can be constructed—the state would capture the possible events that can be executed on the GUI at any instant; transitions cause state changes whenever the number and type of available events changes. Since such applications are also being developed using agile techniques, software testing research must develop new agile mechanisms to test them.

CONCLUSION

This chapter outlined the need for an agile model-based testing mechanism to keep up with agile evolution of software. The example of GUI-based applications was used throughout the chapter. Three existing GUI testing techniques were combined to develop a new process for model-based agile GUI testing. The new process was novel in that it consisted of three iterative sub-processes, each with specific testing goals, resource demands, and tester involvement. The model used to realize this process was discussed.

Each of the three presented techniques has been evaluated in previously reported work and found to be practical. The crash testing approach has been applied on several open-source applications and used to report previously unknown faults in fielded systems; the entire process required a matter of hours with no human intervention (Xie et al., 2005). The smoke testing technique has also been applied to perform testing of nightly builds of several in-house GUI systems (Memon et al., 2005b). The comprehensive testing technique has also been evaluated both for the first time (Memon et al., 2001a) and regression testing (Memon et al., 2003e).

GUIs belong to the wider class of event-driven software. The increasing popularity of event-driven software applications, together with the increased adoption of agile development methodologies fuels the need for the development of other, similar quality assurance techniques for this wider class. The software testing research community needs to understand emerging development trends, and to develop new techniques that leverage the resources available during agile development. The concepts presented in this chapter take the first step towards providing such agile testing techniques.

ACKNOWLEDGMENTS

The authors thank the anonymous reviewers whose comments played an important role in reshaping this chapter. The authors also thank Bin Gan, Adithya Nagarajan, and Ishan Banerjee who helped to lay the foundation for this research. This work was partially supported by the US National Science Foundation under NSF grant CCF-0447864 and the Office of Naval Research grant N00014-05-1-0421.

REFERENCES

Beizer, B. (1990). *Software testing techniques* (2nd ed.). New York: International Thomson Computer Press.

Carloganu, A., & Raguideau, J. (2002). Claire: An event-driven simulation tool for test and validation of software programs. *Proceedings of the 2002 International Conference on Dependable Systems and Networks* (p. 538). Los Alamitos, CA: IEEE Computer Society.

Cheong, E., Liebman, J., Liu, J., & Zhao, F. (2003). Tinygals: A programming model for event-driven embedded systems. *Proceedings of the 2003 ACM symposium on Applied Computing* (pp. 698-704). New York: ACM Press.

Clarke, J. M. (1998). Automated test generation from a behavioral model. *Proceedings of the 11th International Software Quality Week.* San Francisco: Software Research, Inc.

Duarte, S., Martins, J. L., Domingos, H. J., & Preguia, N. (2003). A case study on event dissemination in an active overlay network environment. *Proceedings of the 2nd International Workshop on Distributed Event-Based Systems* (pp. 1-8). New York: ACM Press.

Finsterwalder, M. (2001). Automating acceptance tests for GUI applications in an extreme programming environment. *Proceedings of the 2nd International Conference on eXtreme Programming and Flexible Processes in Software Engineering* (pp. 114-117). New York: Springer.

Gray, J. (2003). What next? A dozen information-technology research goals. *Journal of ACM, 50*(1), 41-57.

Gu, X., & Nahrstedt, K. (2001). An event-driven, user-centric, qos-aware middleware framework for ubiquitous multimedia applications. *Proceedings of the 2001 International Workshop on Multimedia Middleware* (pp. 64-67). New York: ACM Press.

Hicinbothom, J. H., & Zachary, W. W. (1993). A tool for automatically generating transcripts of human-computer interaction. *Proceedings of the Human Factors and Ergonomics Society 37th Annual Meeting, Vol. 2 of SPECIAL SESSIONS: Demonstrations* (pp. 1042). Santa Monica, CA: Human Factors Society.

Holzmann, G. J., & Smith, M. H. (1999). A practical method for verifying event-driven software. *Proceedings of the 21st International Conference on Software Engineering* (pp. 597-607). Los Alamitos, CA: IEEE Computer Society.

Kaddah, M. M. (1993). Interactive scenarios for the development of a user interface prototype. *Proceedings of the 5th International Conference on Human-Computer Interaction, Vol. 2 of I. Software Interfaces* (pp. 128-133). New York: ACM Press.

Kasik, D. J., & George, H. G. (1996). Toward automatic generation of novice user test scripts. *Proceedings of the Conference on Human Factors in Computing Systems: Common Ground* (pp. 244-251). New York: ACM Press.

Kaster, A. (1991). User interface design and evaluation: Application of the rapid prototyping tool EMSIG. *Proceedings of the 4th International Conference on Human-Computer Interaction, Vol. 1 of Congress II: Design and Implementation of Interactive Systems: USABILITY EVALUATION; Techniques for Usability Evaluation* (pp. 635-639). New York: ACM Press.

Lifschitz, V. (1986). On the semantics of STRIPS. In M. P. Georgeff & A. L. Lansky (Eds.), *Reasoning about actions and plans: Proceedings of the 1986 Workshop* (pp. 1-9). San Francisco: Morgan Kaufmann.

Mahajan, R., & Shneiderman, B. (1996). *Visual & textual consistency checking tools for graphical user interfaces.* Technical Report CS-TR-3639, University of Maryland, College Park, May.

Memon, A. M. (2001b). *A comprehensive framework for testing graphical user interfaces (*Doctoral dissertation, University of Pittsburgh, 2001). *Dissertation Abstracts International, 62*, 4084.

Memon, A. M. (2002). GUI testing: Pitfalls and process. *IEEE Computer, 35*(8), 90-91.

Memon, A. M. (2003a). Advances in GUI testing. In *M. V. Zelkowitz (Ed.), Advances in Computers* (Vol. 57). Burlington, MA: Academic Press.

Memon, A. M., & Soffa, M. L. (2003e). Regression testing of GUIs. *Proceedings of the 9th European Software Engineering Conference (ESEC) and 11th ACM SIGSOFT International Symposium on the Foundations of Software Engineering* (pp. 118-127). New York: ACM Press.

Memon, A. M., & Xie, Q. (2004b). Empirical evaluation of the fault-detection effectiveness of smoke regression test cases for GUI-based software. *Proceedings of the International Conference on Software Maintenance 2004 (ICSM'04)* (pp. 8-17). Los Alamitos, CA: IEEE Computer Society.

Memon, A. M., & Xie, Q. (2004c). Using transient/persistent errors to develop automated test oracles for event-driven software. *Proceedings of the International Conference on Automated Software Engineering 2004 (ASE'04)* (pp. 186-195). Los Alamitos, CA: IEEE Computer Society.

Memon, A. M., & Xie, Q. (2005b). Studying the fault-detection effectiveness of GUI test cases for rapidly evolving software. *IEEE Transactions on Software Engineering, 31*(10), 884-896.

Memon, A. M., Banerjee, I., Hashish, N., & Nagarajan, A. (2003b). DART: A framework for regression testing nightly/daily builds of GUI applications. *Proceedings of the International Conference on Software Maintenance* (pp. 410-419). Los Alamitos, CA: IEEE Computer Society.

Memon, A. M., Banerjee, I., & Nagarajan, A. (2003c). GUI ripping: Reverse engineering of graphical user interfaces for testing. *Proceedings of the 10th Working Conference on Reverse Engineering* (pp. 260-269). Los Alamitos, CA: IEEE Computer Society.

Memon, A. M., Banerjee, I., & Nagarajan, A. (2003d). What test oracle should I use for effective GUI testing? *Proceedings of the IEEE International Conference on Automated Software Engineering* (pp. 164-173). Los Alamitos, CA: IEEE Computer Society.

Memon, A., Nagarajan, A., & Xie, Q. (2005a). Automating regression testing for evolving GUI software. *Journal of Software Maintenance and Evolution: Research and Practice, 17*(1), 27-64.

Memon, A. M., Pollack, M. E., & Soffa, M. L. (1999). Using a goal-driven approach to generate test cases for GUIs. *Proceedings of the 21st International Conference on Software Engineering* (pp. 257-266). New York: ACM Press.

Memon, A. M., Pollack, M. E., & Soffa, M. L. (2000a). Automated test oracles for GUIs. *Proceedings of the ACM SIGSOFT 8th International Symposium on the Foundations of Software Engineering* (pp. 30-39). New York: ACM Press.

Memon, A. M., Pollack, M. E., & Soffa, M. L. (2000b). Plan generation for GUI testing. *Proceedings of the 5th International Conference on Artificial Intelligence Planning and Scheduling* (pp. 226-235). Menlo Park, CA: AAAI Press.

Memon, A. M., Pollack, M. E., & Soffa, M. L. (2001a). Hierarchical GUI test case generation using automated planning. *IEEE Transactions on Software Engineering, 27*(2), 144-155.

Memon, A. M., Porter, A., Yilmaz, C., Nagarajan, A., Schmidt, D. C., & Natarajan, B. (2004a). Skoll: distributed continuous quality assurance. *Proceedings of the 26th IEEE/ACM International Conference on Software Engineering* (pp. 459-468). New York: ACM Press.

Memon, A. M., Soffa, M. L., & Pollack, M. E. (2001c). Coverage criteria for GUI testing. *Proceedings of the 8th European Software Engineering Conference (ESEC) and 9th ACM SIGSOFT International Symposium on the Foundations of Software Engineering (FSE-9)* (pp. 256-267). New York: ACM Press.

Mulligan, R. M., Altom, M. W., & Simkin, D. K. (1991). User interface design in the trenches: Some tips on shooting from the hip. *Proceedings of ACM CHI'91 Conference on Human Factors in Computing Systems, Practical Design Methods* (pp. 232-236). New York: ACM Press.

Myers, B. A. (1993a). *Why are human-computer interfaces difficult to design and implement?* Technical Report CS-93-183, Carnegie Mellon University, School of Computer Science.

Myers, B. A. (1995a). *State of the art in user interface software tools. Human-computer interaction: Toward the year 2000.* San Francisco: Morgan Kaufmann Publishing.

Myers, B. A. (1995b). User interface software tools. *ACM Transactions on Computer-Human Interaction, 2*(1), 64-103.

Myers, B. A., & Olsen, JR., D. R. (1994). User interface tools. *Proceedings of ACM CHI'94 Conference on Human Factors in Computing Systems, TUTORIALS* (Vol. 2, pp. 421-422). New York: ACM Press.

Myers, B. A., Hollan, J. D., & Cruz, I. F. (1996). Strategic directions in human-computer interaction. *ACM Computing Surveys, 28*(4), 794-809.

Myers, B. A., Olsen, JR., D. R., & Bonar, J. G. (1993b). User interface tools. *Proceedings of ACM INTERCHI'93 Conference on Human Factors in Computing Systems: Adjunct Proceedings, Tutorials* (p. 239). New York: ACM Press.

Nielsen, J. (1993). Iterative user-interface design. *IEEE Computer, 26(*11), 32-41.

Pednault, E. P. D. (1989). ADL: Exploring thc middle ground between STRIPS and the situation calculus. *Proceedings of KR'89* (pp. 324-331). San Francisco: Morgan Kaufmann Publisher.

Perry, W. (1995). *Effective methods for software testing.* New York: John Wiley & Sons.

Rothermel, G., & Harrold, M. J. (1997). A safe, efficient regression test selection technique. *ACM Transactions on Software Engineering and Methodology, 6*(2), 173-210.

Shehady, R. K., & Siewiorek, D. P. (1997). A method to automate user interface testing using variable finite state machines. *Proceedings of the 27th Annual International Conference on Fault-Tolerant Computing* (pp. 80-88). Los Alamitos, CA: IEEE Computer Society.

Sliwa, C. (2003). Event-driven architecture poised for wide adoption. *COMPUTERWORLD.* Retrieved May 12, 2005, from http://www.computerworld.com/softwaretopics/software/appdev/story/0,10801,81133,00.html

Weld, D. S. (1994). An introduction to least commitment planning. *AI Magazine, 15*(4), 27-61.

Weld, D. S. (1999). Recent advances in AI planning. *AI Magazine, 20*(1), 55-64.

Welsh, M., Culler, D., & Brewer, E. (2001). Seda: An architecture for well-conditioned, scalable internet services. *Proceedings of the 18th ACM Symposium on Operating Systems Principles* (pp. 230-243). New York: ACM Press.

White, L. (1996). Regression testing of GUI event interactions. *Proceedings of the International Conference on Software Maintenance (ICSM'96)* (pp. 350-358). Los Alamitos, CA: IEEE Computer Society.

White, L., Almezen, H., & Alzeidi, N. (2001). User-based testing of GUI sequences and their interactions. *Proceedings of the 12th International Symposium Software Reliability Engineering* (pp. 54-63). Los Alamitos, CA: IEEE Computer Society.

Wick, D. T., Shehad, N. M., & Hajare, A. R. (1993). Testing the human computer interface for the telerobotic assembly of the space station. *Proceedings of the 5th International Conference on Human-Computer Interaction, Special Applications* (Vol. 1, pp. 213-218). New York: ACM Press.

Xie, Q., & Memon, A. M. (2005). Rapid crash testing for continuously evolving GUI-based software applications. *Proceedings of the International Conference on Software Maintenance 2005* (pp. 473-482). Los Alamitos, CA: IEEE Computer Society.

Section III
Quality within Agile Process Management

Chapter VII
Software Configuration Management in Agile Development

Lars Bendix
Lund Institute of Technology, Sweden

Torbjörn Ekman
Lund Institute of Technology, Sweden

ABSTRACT

Software configuration management (SCM) is an essential part of any software project and its importance is even greater on agile projects because of the frequency of changes. In this chapter, we argue that SCM needs to be done differently and cover more aspects on agile projects. We also explain how SCM processes and tools contribute both directly and indirectly to quality assurance. We give a brief introduction to central SCM principles and define a number of typical agile activities related to SCM. Subsequently, we show that there are general SCM guidelines for how to support and strengthen these typical agile activities. Furthermore, we establish a set of requirements that an agile method must satisfy to benefit the most from SCM. Following our general guidelines, an agile project can adapt the SCM processes and tools to its specific agile method and its particular context.

INTRODUCTION

In traditional software development organisations, software configuration management (SCM) is often pushed onto the projects by the quality assurance (QA) organisation. This is done because SCM in part can implement some QA measures and in part can support the developers in their work and therefore helps them to produce better quality. The same holds true for agile methods—SCM can directly and in-directly contribute to better QA on agile projects.

Software configuration management (SCM) is a set of processes for managing changes and modifications to software systems during their entire life cycle. Agile methods embrace change and focus on how to respond rapidly to changes in the requirements and the environment (Beck, 1999a). So it seems obvious that SCM should be an even more important part of agile methods

Figure 1. The different layers of SCM

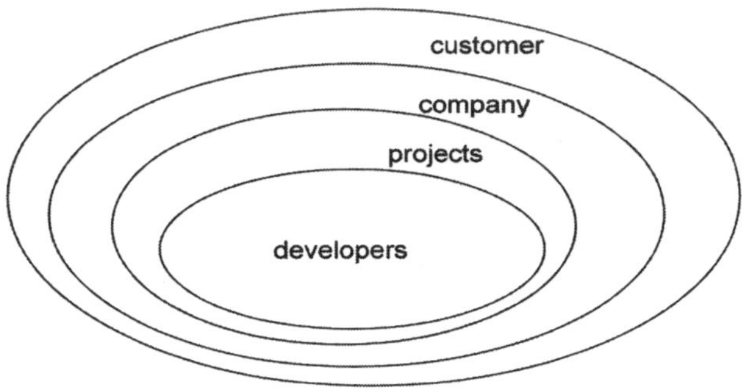

than it is of traditional development methods. However, SCM is often associated with heavily process-oriented software development and the way it is commonly carried out might not transfer directly to an agile setting. We believe there is a need for SCM in agile development but that ist should be carried out in a different way. There is a need for the general values and principles of SCM, which we consider universal for all development methods, and there is a need for the general techniques and processes, which we are certain will be of even greater help to agile developers than they are to traditional developers.

There are some major differences in agile projects compared to traditional projects that heavily influence the way SCM can—and should—be carried out. Agile methods shift the focus from the relation between a project's management and the customer to the relation between developers and the customer. While traditional SCM focuses on the projects and company layers in Figure 1, there is a need to support developers as well when using SCM in agile projects. Shorter iterations, more frequent releases, and closer collaboration within a development team contribute to a much greater stress on SCM processes and tools.

Agile methods are people-oriented rather than process-oriented and put the developer and the customer in focus. As a consequence, SCM has to shift its primary focus from control activities to that of service and support activities. The main focus on audits and control needs to be replaced by a main focus on supporting the highly iterative way of working of both the team and the developers, as seen in Figure 2. From a QA point of view, the control measures are moved down to the developers themselves with the purpose of shortening the feedback loop in agile methods. So

Figure 2. The main development loops in agile

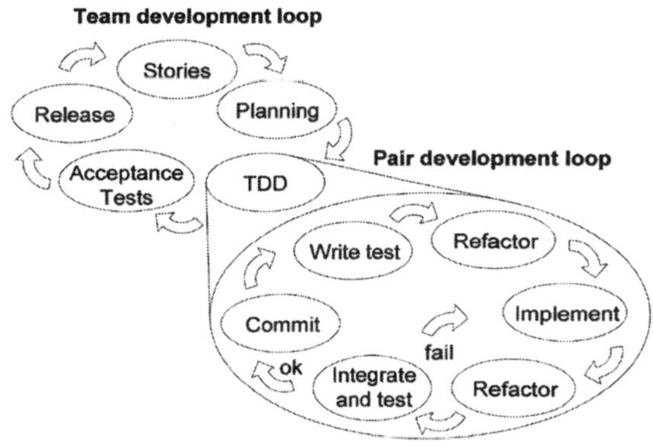

SCM does not directly contribute to the QA on an agile project, this is the task of the processes that the agile method in question prescribes. However, by supporting said processes and making them easier and safer to practice SCM indirectly is a big factor in QA on agile projects.

The traditional process-oriented view of SCM has also lead to several misconceptions of agile methods from an SCM point of view. The lack of explicit use of SCM and its terminology has lead quite a few people to conclude that agile methods are not safe due to an apparent lack of rigorous change management. However, a lot of SCM activities are actually carried out in agile methods although they are not mentioned explicitly. Bendix and Hedin (2002) identify a need for better support from SCM, in particular for refactoring in order for this practice to be viable. Koskela (2003) reviews agile methods in general from an SCM point of view and concludes that only a few of the existing agile methods take SCM explicitly into account. He also notices that most methods highly value SCM tool support but that SCM planning has been completely forgotten. There is thus a need to provide guidance for using SCM or for implementing SCM in agile. The SCM literature mostly takes the control-oriented view of SCM (Berlack, 1992; Buckley, 1993; Hass, 2003; Leon, 2005) and there is very little written about team- and developer-oriented support activities (Babich, 1986; Mikkelsen & Pherigo, 1997; Bendix & Vinter, 2001; Berczuk & Appleton, 2003). These activities are the ones that can benefit agile methods the most and should therefore be emphasized more when used in an agile setting. However, it is important to stress that agile methods need the whole range of SCM support from developer through to customer.

In the next section, we provide background information about SCM for those who are not so familiar with SCM, and describe and define a number of SCM-related agile activities to establish a terminology. In Section 3, we give general guidelines for how these agile activities can be supported by SCM and how agile methods could benefit from adopting more SCM principles. We also provide pointers to literature where more details can be found. Future trends for SCM in the agile context are described in Section 4, and in Section 5 we draw our conclusions.

BACKGROUND

This section gives an introduction to the concepts and terminology in SCM that serve as a background for the analysis in following sections. We also define and describe activities in agile methods that are related to SCM or affected by SCM in one way or the other.

SCM Activities

SCM is a method for controlling the development and modifications of software systems and products during their entire life cycle (Crnkovic, Asklund, & Persson Dahlqvist, 2003). From this viewpoint, SCM is traditionally divided into the following activities: configuration identification, configuration control, configuration status accounting, and configuration audit (Leon, 2005). These activities reflect mostly the part of a development project with relations to the customer. However, since agile methods are often more developer centric, there is also a need for a more developer-oriented view of SCM than the traditional control-oriented view above. Typical developer-oriented aspects of SCM include: version control, build management, workspace management, concurrency control, change management, and release management (Bendix & Vinter, 2001). We present each activity from a general perspective and explain both its purpose and what is included in the activity. After this introduction, the reader should be familiar with these basic SCM concepts and their purpose, so we can use them for our analysis in the next section.

Configuration Identification

Configuration identification is the activity where a system is divided into uniquely identifiable components, called configuration items, for the purpose of software configuration management. The physical and functional characteristics of each configuration item are documented including its interfaces and change history. Each configuration item is given a unique identifier and version to distinguish it from other items and other versions of the same item. This allows us to reason about a system in a consistent way both regarding its structure and history. Each item can be either a single unit or a collection (configuration) of lower level items allowing hierarchical composition. During configuration identification a project baseline and its contents are also defined, which helps to control change as all changes apply to this uniquely defined baseline.

Configuration Control

Software is very different from hardware as it can be changed quickly and easily, but doing so in an uncontrolled manner often leads to chaos. Configuration control is about enabling this flexibility in a controlled way through formal change control procedures including the following steps: evaluation, coordination, approval or disapproval, and implementation of changes. A proposed change request typically originates from requests for new features, enhancement of existing features, bug reports, etc. A request is first evaluated by a Change Control Board (CCB) that approves or disapproves the request. An impact analysis is performed by the CCB to determine how the change would affect the system if implemented. If a request is approved, the proposed change is assigned to a developer for implementation. This implementation then needs to be verified through testing to ensure that the change has been implemented as agreed upon before the CCB can finally close the change request.

Configuration Status Accounting

Developers are able to track the current status of changes by formalizing the recording and reporting of established configuration items, status of proposed changes, and implementation of approved changes. Configuration status accounting is the task to provide all kinds of information related to configuration items and the activities that affect them. This also includes change logs, progress reports, and transaction logs. Configuration status accounting enables tracking of the complete history of a software product at any time during its life cycle and also allows changes to be tracked compared to a particular baseline.

Configuration Audits

The process of determining whether a configuration item, for instance a release, conforms to its configuration documents is called configuration audit. There are several kinds of audits each with its own purpose but with the common goal to ensure that development plans and processes are followed. A functional configuration audit is a formal evaluation that a configuration item has achieved the performance characteristics and functions defined in its configuration document. This process often involves testing of various kinds. A physical configuration audit determines the conformity between the actual produced configuration item and the configuration according to the configuration documents. A typical example is to ensure that all items identified during configuration identification are included in a product baseline prior to shipping. An in-process audit ensures that the defined SCM activities are being properly applied and controlled and is typically carried out by a QA team.

Version Control

A version control system is an invaluable tool in providing history tracking for configuration items. Items are stored, versions are created, and their historical development is registered and conveniently accessible. A fundamental invariant is that versions are immutable. This means that as soon as a configuration item is given a version number, we are assured that it is unique and its contents cannot be changed unless we create a new version. We can therefore recreate any version at any point in time. Version control systems typically support configuration status accounting by providing automatic support for history tracking of configuration items. Furthermore, changes between individual versions of a configuration item can be compared automatically and various logs are typically attached to versions of a configuration item.

Build Management

Build management handles the problem of putting together modules in order to build a running system. The description of dependencies and information about how to compile items are given in a system model, which is used to derive object code and to link it together. Multiple variants of the same system can be described in a single system model and the build management tool will derive different configurations, effectively building a tailored system for each platform or product variant. The build process is most often automated, ranging from simple build scripts to compilation in heterogeneous environments with support for parallel compilation. Incremental builds, that only compile and link what has changed, can be used during development for fast turn around times, while a full build, rebuilding the entire system from scratch, is normally used during system integration and release.

Workspace Management

The different versions of configuration items in a project are usually kept in a repository by the version control tool. Because these versions must be immutable, developers cannot be allowed to work directly within this repository. Instead, they have to take out a copy, modify it, and add the modified copy to the repository. This also allows developers to work in a controlled environment where they are protected from other people's changes and where they can test their own changes prior to releasing them to the repository. Workspace management must provide functionality to create a workspace from a selected set of files in the repository. At the termination of that workspace, all changes performed in the workspace need to be added to the repository. While working in the workspace, a developer needs to update his workspace, in a controlled fashion, with changes that other people may have added to the repository.

Concurrency Control

When multiple developers work on the same system at the same time, they need a way to synchronize their work; otherwise it may happen that more than one developer make changes to the same set of files or modules. If this situation is not detected or avoided, the last developer to add his or her changes to the repository will effectively erase the changes made by others. The standard way to avoid this situation is to provide a locking mechanism, such that only the developer who has the lock can change the file. A more flexible solution is to allow people to work in parallel and then to provide a merge facility that can combine changes made to the same file. Compatible changes can be merged automatically while incompatible changes will result in a merge conflict that has to be resolved manually. It is worth noticing that conflicts are resolved in the workspace of the developer that triggered the conflict, who is the proper person to resolve it.

Change Management

There are multiple and complex reasons for changes and change management needs to cover all types of changes to a system. Change management includes tools and processes that support the organization and tracking of changes from the origin of the change to the approval of the implemented source code. Various tools are used to collect data during the process of handling a change request. It is important to keep traceability between a change request and its actual implementation, but also to allow each piece of code to be associated to an explicit change request. Change management is also used to provide valuable metrics about the progress of project execution.

Release Management

Release management deals with both the formal aspects of the company releasing to the customer and the more informal aspects of the developers releasing to the project. For a customer release, we need to carry out both a physical and a functional configuration audit before the actual release. In order to be able to later re-create a release, we can use a bill-of-material that records what went into the release and how it was built. Releasing changes to the project is a matter of how to integrate changes from the developers. We need to decide on when and how that is done, and in particular on the "quality" of the changes before they may be released.

Agile Activities

This section identifies a set of agile activities that either implement SCM activities or are directly affected by SCM activities. The presentation builds on our view of agile methods as being incremental, cooperative, and adaptive. Incremental in that they stress continuous delivery with short release cycles. Cooperative in that they rely on teams of motivated individuals working towards a common goal. Adaptive in that they welcome changing requirements and reflect on how to become more effective. While all activities presented in this section may not be available in every agile method, we consider them representative for the agile way of developing software.

Parallel Work

Most projects contain some kind of parallel work, either by partitioning a project into sub-projects that are developed in parallel, or by implementing multiple features in parallel.

Traditional projects often try to split projects into disjoint sub-projects that are later combined into a whole. The incremental and adaptive nature of agile methods requires integration to be done continuously since new features are added as their need is discovered. Agile methods will therefore inevitably lead to cooperative work on the same, shared code base, which needs to be coordinated. To become effective, the developers need support to work on new features in isolation and then merge their features into the shared code base.

Continuous Integration

Continuous integration means that members of a team integrate their changes frequently. This allows all developers to benefit from a change as soon as possible, and enables early testing of changes in their real context. Continuous integration also implies that each member should integrate changes from the rest of the team for early detection of incompatible changes. The frequent integration decreases the overall integration cost since incompatible changes are detected and resolved early, in turn reducing the complex integration problems that are common in traditional projects that integrate less often.

Regular Builds

Agile projects value frequent releases of software to the customer and rapid feedback. This implies more frequent builds than in traditional projects. Releases, providing added value to the customer, need to be built regularly, perhaps on a weekly or monthly basis. Internal builds, used by the team only, have to be extremely quick to enable rapid feedback during continuous integration and test-driven development. This requires builds to be automated to a large extent to be feasible in practice.

Refactoring

The incremental nature of agile methods requires continuous Refactoring of code to maintain high quality. Refactorings need to be carried out as a series of steps that are reversible, so one can always back out if a refactoring does not work. This practice relies heavily on automated testing to ensure that a change does not break the system. In practice, this also means that it requires quick builds when verifying behavioural preservation of each step.

Test-Driven Development

Test-driven development is the practice that test drives the design and implementation of new features. Implementation of tests and production code is interleaved to provide rapid feedback on implementation and design decisions. Automated testing builds a foundation for many of the presented practices and requires extremely quick builds to enable a short feedback loop.

Planning Game

The planning game handles scheduling of an XP project. While not all agile methods have an explicit planning game, they surely have some kind of lightweight iterative planning. We emphasize planning activities such as what features to implement, how to manage changes, and how to assign team resources. This kind of planning shares many characteristics with the handling of change requests in traditional projects.

SCM IN AN AGILE CONTEXT

In the previous section, we defined some agile activities that are related to SCM and we also outlined and described the activities that make up the field of SCM. In this section, we will show how SCM can provide support for such agile activities so they succeed and also how agile methods can gain even more value from SCM. It was demonstrated in Asklund, Bendix, and Ekman (2004) that agile methods, in this case exemplified by XP, do not go against the fundamental principles of SCM. However, it also showed that, in general, agile methods do not provide explicit nor complete guidance for using or implementing SCM. Furthermore, the focus of SCM also needs to shift from control to service and support (Angstadt, 2000) when used in agile. SCM does not require compliance from agile, but has a lot of good advice that you can adapt to your project if you feel the need for it—and thus value people and interactions before tools and processes (Agile Manifesto, 2001).

In this section, we first look at how SCM can support and service the agile activities we defined in the previous section. Next, we look at how agile methods could add new activities and processes from SCM and in this way obtain the full benefit of support from SCM.

How Can SCM Techniques Support Agile?

SCM is not just about control and stopping changes. It actually provides a whole range of techniques and processes that can service and support also agile development teams. Agile methods may tell you what you should do in order to be agile or lean, but in most cases, they are also very lean in actually giving advice on how to carry out these agile processes. In this sub-section, we show how SCM techniques can be used to support and strengthen the following SCM-related agile activities: parallel work, continuous integration, regular builds, refactoring, test-driven development, and planning game.

Parallel Work

Agile teams will be working in parallel on the same system. Not only on different parts of the system leading to shared data, but also on the same parts of the system, leading to simultaneous update and double maintenance. Babich (1986) explains all the possible problems there are when coordinating a team working in parallel—and also the solutions.

The most common way of letting people work in parallel is not to have collective code ownership, but private code ownership and locking of files that need to be changed. This leads to a "split and combine" strategy where only one person owns some specific code and is allowed to change it. Industry believes that this solves the problem, but the "shared data" problem (Babich, 1986) shows that even this apparently safe practice has its problems (e.g., combining the splits). These problems are obviously present if you practise parallel work as well. In addition, we have to solve the "simultaneous update" problem and the "double maintenance" problem, when people actually work on the same file(s) in parallel.

The "shared data" problem is fairly simple to solve—if the problem is sharing, then isolate yourself. Create a physical or virtual workspace that contains all of the code and use that to work in splendid isolation from other people's changes. Obviously you cannot ignore that other people make changes, but having your own workspace, you are in command of when to "take in" those changes and will be perfectly aware of what is happening.

The "simultaneous update" problem only occurs for collective code ownership where more people make changes to the same code at the same time. Again, the solution is fairly simple, you must be able to detect that the latest version, commonly found in the central repository, is not the version that you used for making your changes. If that is not the case, it means that someone has worked in parallel and has put a new version into the repository. If you add your version to the repository, it will "shadow" the previous version and effectively undo the changes done in that version. If you do not have versioning, the new version will simply overwrite and permanently erase the other person's changes. Instead you must "integrate" the parallel changes and put the resulting combined change into the repository or file system. There are tools that can help you in performing this merge.

The "double maintenance" problem is a consequence of the "protection" from the "shared data" problem. In the multiple workspaces, we will have multiple copies of every file and according to Babich (1986) they will soon cease to be identical. When we make a change to a file in one workspace, we will have to make the same change to the same file in all the other workspaces to keep the file identical in all copies. It sounds complicated but is really simple, even though it requires some discipline. Once you have made a change, you put it in the repository and—sooner or later—the other people will take it in from the repository and integrate it if they have made changes in parallel (see the "simultaneous update" problem).

A special case of parallel work is distributed development where the developers are physically separated. This situation is well known in the SCM community and the described solutions (Bellagio & Milligan 2005) are equally applicable to distributed development as to parallel work. There are solutions that make tools scale to this setting as well. Distributed development is thus not different from parallel work from an SCM perspective, as long as the development process that SCM supports scales to distributed development.

In summary, we need a repository where we can store all the shared files and a workspace where we can change the files. The most important aspect of the repository is that it can detect parallel work and that it can help us in sorting out such parallel work. Also it should be easy and simple to create whole workspaces. Most version control tools are able to do that and there is no need to use locking, which prevents real parallel work, since optimistic sharing works well. We must thus choose a tool that can implement the copy-merge work model (Feiler, 1991).

Continuous Integration

In traditional projects, the integration of the contributions of many people is always a painful process that can take days or even weeks. Therefore, continuous integration seems like a mission impossible, but this is actually not the case. The reason why integration is painful can be found in the "double maintenance" problem (Babich, 1986)—the longer we carry on the double maintenance without integrating changes, the greater the task of integration will be. So there are good reasons for integrating as often as possible, for instance after each added feature.

Integrating your change into the team's shared repository is often a two-step process. The reason is that tools usually cannot solve merge conflicts and re-run automated tests to check the quality in one step. First, you have to carry out a "download" (or subscription) integration where you take all the changes that have been added to the repository since you last integrated and integrate them into your workspace, as shown in Figure 3, where a box represents a new version of the configuration. If nothing new has happened, you are safe and can do the "upload" (or publication) integration, which simply adds your changes as the latest versions in the repository. If something has changed in the repository, it can be either new versions of files that you have not changed—these can simply be copied into your workspace—or files that you have changed where there may be conflicting changes. In the latter case you have to merge the repository changes into your own changes. At this point, all other people's changes have been integrated with your changes and your workspace is up-to-date, so you could just add the result to the repository. However, you should check that the integration actually produced a viable result and check the quality of it. This can be done by running a set of quality tests (e.g., unit tests, acceptance tests), and if everything works well, then you can add the result to the repository—if your workspace is still up-to-date. Otherwise, you have to continue to do "download" integrations and quality checks until you finally succeed and can do the "upload" integration, as shown in Figure 3.

This way of working (except for the upload quality control) is implemented in the strict long transactions work model (Feiler, 1991). You will notice that in this process, the upload integration is a simple copy of a consistent and quality assured workspace. All the work is performed in the download integration. Following the advice of Babich (1986), this burden can be lessened if it is carried out often as the changes you have to integrate are smaller. So for your own sake you should download integrate as often as possible.

Figure 3. Download and upload integration

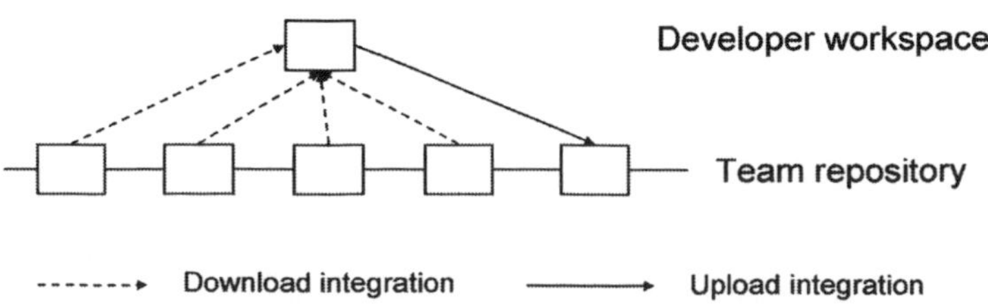

Moreover, for the sake of the team you should upload (publish) immediately when you have finished a task or story so other people get the possibility to synchronize their work with yours.

What we have described here is the commonality between the slightly different models and approaches presented in Aiello (2003), Appleton, Berczuk, and Konieczka (2003a, 2003b, 2004a), Appleton, Berczuk, and Cowham (2005), Farah (2004), Fowler and Foemmel (2006), Moreira (2004) and Sayko (2004). If you are interested in the details about how you can vary your approach to continuous integration depending on your context, you can consult the references.

Continuous integration leads to an increased velocity of change compared to traditional development. This puts additional strains on the integration process but is not a performance problem on the actual integration per se. However, there may be performance issues when the integration is combined with a quality gate mechanism used to determine whether changes are of sufficient quality to be integrated in the common repository or not. Even if this quality gate process is fully automated, it will be much slower than the actual merge and upload operation and may become a bottleneck in the integration process. It may therefore not always be possible to be true to the ideal that developers should carefully test their code before uploading their changes in which case you could use a more complex model for continuous integration (Fowler & Foemmel, 2006) that we will describe next under regular builds.

Regular Builds

When releases become frequent it also becomes important to be able to build and release in a lean way. If not, much time will be "wasted" in producing these releases that are needed to get customer feedback. Making it lean can be done in three ways: having always releasable code in the repository, performing a less formal release process, and automation of the build and release processes.

Before you can even think about releasing your code, you have to assure that the code you have is of good quality. In traditional development methods this is often done by a separate team that integrates the code and does QA. In agile, this is done by the developers as they go. The ideal situation is that the code in the repository is always of the highest quality and releasable at any time. This is not always possible and you can then use a mix between the traditional way and the agile ideal by having multiple development lines. The developers use an integration line to check in high quality code and to stay in sync with the rest of the developers. The QA-team uses a separate line to pull in changes from the integration line and does a proper and formal QA before they "promote" the approved change to the release line, as seen in Figure 4.

In agile methods, there is a general tendency to move the focus of QA from coming late in the development process, just before release, to being a centre of attention as early as possible in the development process. This means that agile can do with a less formal release process than traditional projects because much of the work has already been done. However, there is still a need to do physical and functional audits and to work with bill-of-materials such that earlier releases can be re-created again if needed. In agile methods, functional audits can be carried out by running the acceptance tests. They are the specification of the requirements that should be implemented. To be really sure that we have implemented everything we claim, we should check the list of acceptance tests against the list of requirements we claim have been implemented in this release. We also need to check whether all the files that should be in the release (e.g., configuration files, manual, documentation, etc.) are actually there.

When releasing becomes a frequent action, there is a much greater need to automate it. The

Figure 4. Working with integration and release lines

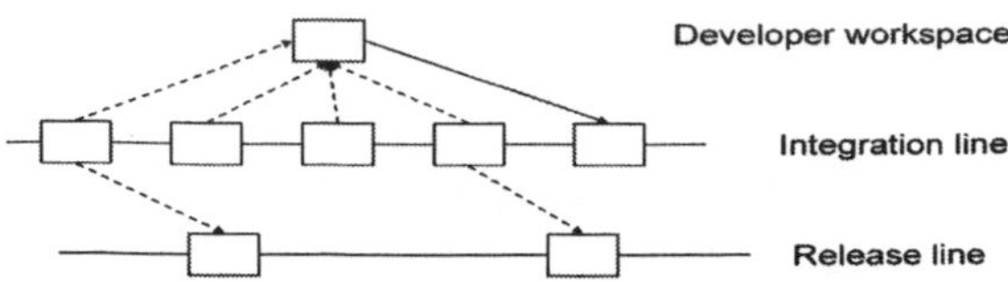

actual creation of the release can be automated by using build tools; acceptance tests and the verification that all files are there can be automated by writing simple scripts.

More information about regular builds can be found in Appleton and Cowham (2004b).

Refactoring

Refactoring is an important part of agile methods but also to some extent in traditional methods. The purpose of a Refactoring is not to implement new functionality, but rather to simplify the code and design.

In general, there are two different situations where you do refactorings: as part of a story to simplify the code before and/or after the implementation of the story's functionality; and architectural refactorings that are needed to implement a whole new set of features. In both cases, the two main problems are that a refactoring may touch large parts of the code and that the refactoring should be traceable and possible to undo. The latter means that there is the need for version control tool to keep track of the steps of each refactoring and make it possible to back out of a refactoring if it turns out that is does not work.

The fact that refactorings tend to be "global" possibly affecting large parts of the code, puts even greater strains on the continuous integration since there are more possibilities of merge conflicts. The recommendation for successful application of continuous integration is to integrate very often to reduce the risk of merge conflicts. The same goes for refactorings that should be split up into many small steps that are integrated immediately when they are done.

If you need to refactor code to facilitate the ease of implementing a story, then this refactoring should be seen as a separate step and integrated separately—the same goes if you need to refactor after the implementation of the story. For the architectural refactorings, we need to split the refactoring up into smaller tasks such that there will be as little time as possible between integrations to lower the risk of merge conflicts. Larger refactorings should also be planned and analysed for impact such that it is possible to coordinate the work to keep down the parallel work, or at least to make people aware of the fact that it is going on.

For a more specific treatment of the problems architectural refactorings can cause to SCM tools and the continuous integration process and how these problems can be dealt with, we refer the reader to Ekman and Asklund (2004) and Dig, Nguyen, and Johnson (2006).

Test-Driven Development

The short version of test-driven development is design a little—where you design and write tests, code a little, and finally run the tests to get feedback. Here the crucial part is to get feedback on what you have just changed or added. If that cannot happen very quickly, test-driven development breaks down with respect to doing it in small increments. If you want to run your tests after writing a little code, you must be able to re-build the application you want to test very quickly—if you have to wait too long you are tempted to not follow the process as it is intended.

So what is needed is extremely quick re-builds, a matter of a few minutes or less, and the good news is that SCM can provide that. There are techniques for doing minimal, incremental builds that will give you a fast turn-around time, so you can run your tests often without having to wait too long. Make (Feldman, 1979) is the ancestor of all minimal, incremental build tools, but there exists a lot of research on how to trade "consistency" of a build for time (Adams, Weinert, & Tichy, 1989; Schwanke & Kaiser, 1988). For the small pair development loop in Figure 2, we might be satisfied with less than 100% consistency of the build as long as it is blisteringly fast. For the big team development loop in Figure 2 (i.e., integrating with others), speed might not be that important while consistency of the build is crucial. A properly set up SCM system will allow developers to have flexible build strategies that are tailored to specific parts of their development cycle.

Another aspect of test-driven development is that if we get an unexpected result of a test-run, then we have to go bug hunting. What is it that has caused the malfunction? If you run tests often, it means that you introduced the bug in the code that you wrote most recently—or as Babich puts it "an ounce of derivation is worth a pound of analysis" (Babich, 1986)—meaning that if we can tell the difference in the code between now and before, we are well under way with finding the bug. Version control tools provide functionality for showing the difference between two versions of the same file and some tools can even show the structural differences between two versions of a configuration.

Planning Game

Agile methods use stories, or similar lightweight specification techniques, as the way that customers specify the requirements of the system, and acceptance tests to specify the detailed functionality. These stories specify changes to the system and correspond to change requests when analyzed from an SCM perspective. The stories, or change requests, have to be estimated for implementation cost by the developers and then prioritised and scheduled by the customer during the planning game. For someone coming from SCM this sounds very much like the traditional way of handling change requests: an impact analysis has to be carried out to provide sufficient information for the Change Control Board to be able to make its decision whether to implement the change request, defer it, or reject it. So we can see that the parallel to estimation in agile is impact analysis (Bohner & Arnold, 1996) in traditional SCM. Likewise, the parallel to the customer prioritising is the chair of the Change Control Board taking decisions (Daniels, 1985). For the planning game to work properly, it is important that everyone is aware of what his or her role is—and that they seek information that will allow them to fill that role well. It is also important to be aware of the fact that the traditional formal change request handling process can indeed—and should—be scaled to fit the agility and informality that is needed in an agile method.

How Can SCM Add More Value to Agile?

In agile methods, there is very much focus on the developers and the production process. In the previous sub-section, we have seen how many of these processes can be well supported by techniques and principles from SCM. However, agile methods often overlook the aspects of SCM that deal with the relation to the customer and where traditional SCM has special emphasis. In the following, we look at the traditional SCM as represented by the four activities of configuration identification, configuration control, configuration status accounting, and configuration audit (Leon, 2005). For each activity, we describe what new activities and processes could be added to agile methods to help provide a more covering support for the development team.

Configuration Identification

The most relevant part of configuration identification for agile methods is the identification and organisation of configuration items. Some artefacts are so important for a project that they become configuration items and go into the shared repository. Other artefacts (e.g., sketches, experiments, notes, etc.) have a more private nature and they should not be shared in order not to confuse other people. However, it may still be convenient to save and version some of the private artefacts to benefit from versioning even though they are not configuration items. They can be put into the repository but it is very important that the artefacts, configuration items and not, are structured in such a way that it is absolutely clear what a configuration item is and what a private artefact is. Structuring of the repository is an activity that is also important when it contains only configuration items.

Configuration identification is an SCM activity that traditionally is done up-front, which goes against the agile philosophy. However, there can be some reason in actually trying to follow the experience that SCM provides. Rules for identifying configuration items should be agreed upon, such that they can be put into the repository and information about them shared as early as possible. More importantly, though, is that the structuring of configuration items should not be allowed to just grow as the project proceeds, because most SCM tools do not support name space versioning (Milligan, 2003) (i.e., handling structural changes to the repository while retaining the change history).

Configuration Control

The part of configuration control that deals with the handling of change requests is taken care of by a planning game or similar activity. However, two important aspects of configuration control are neglected by most agile methods: tracking and traceability.

In traditional SCM, change requests are tracked through their entire lifetime from conception to completion. At any given point in time, it is important to know the current state of the change request and who has been assigned responsibility for it. This can benefit agile methods too as they also need to manage changes and coordinate the work of different people. In some agile methods there is an explicit tracker role (chromatic, 2003) that is responsible for this activity.

Traceability is an important property of traditional SCM and is sometimes claimed to be the main reason for having SCM. It should be possible to trace changes made to a file back to the specific change request they implement. Likewise, it should be possible to trace the files that were changed when implementing a certain change request. The files that are related to a change request are not just source code files, but all files that are affected by that change (e.g., test cases, documentation, etc). Another aspect of traceability is to be able to know exactly what went into a specific build or release—and what configurations contain a specific version of a specific file. The main advantage of having good traceability is to allow for a better impact analysis so we can be informed of the consequences of changes and improve the coordination between people on the team.

Configuration Status Accounting

This activity should be seen as a service to everyone involved in a project including developers and the customer, even though it traditionally has been used primarily by management and in particular project managers. Configuration status accounting can be looked at as simple data mining where you collect and present data of interest. Many agile methods are very code centred and the repository is the place where we keep the configuration items that are important for the project, so it is natural to place the meta-data to mine in the same repository. Configuration status accounting does not need to be an upfront activity like configuration identification, but can be added as you discover the need. However, you should be aware that the later you start collecting data to mine, the less data and history you get to mine. Usually this is seen as an activity that benefits only managers, but there can be much support for the developers too—all you have to do it to say what kind of meta-data you want collected and how you want it to be presented. If you do not document changes in writing, then it is important that you can get hold of the person that did a change; when you have shared code, then it is important to see who is currently working on what.

Configuration Audit

Configuration audit can be looked at as a verification activity. The actual work, considered as a QA activity, has been done elsewhere as part of other processes, but during the configuration audits, it gets verified that it has actually been carried out. The functional configuration audits verify that we have taken care of and properly closed all change requests scheduled for a specific build or release. The physical configuration audit is a "sanity check" that covers the physical aspects (e.g., that all components/files are there—CD, box, manual) and that it can actually be installed. Even though configuration audit is not directly a QA activity, it contributes to the quality of the product by verifying that certain SCM and QA activities have actually been carried out as agreed upon. Configuration audits are needed not because we mistrust people, but because from time to time people can be careless and forget something. The basis for automating the functional configuration audit in agile is there through the use of unit and acceptance tests.

SCM Plans and Roles

You definitely need to plan and design your SCM activities and processes very carefully on an agile project. Moreover, they have to be carried out differently from how they are done on traditional projects and the developers will need to know more about SCM because they are doing more of it on an agile project.

This does not imply that you should write big detailed SCM plans the same way as it is being done for traditional projects. The agile manifesto (Agile Manifesto, 2001) values working software over comprehensive documentation. The same goes for SCM where you should value working SCM processes over comprehensive SCM plans. In general, what needs to be documented are processes and activities that are either complex or carried out rarely. The documentation needs to be kept alive and used—otherwise it will not be up-to-date and should be discarded. We can rely on face-to-face conversation to convey information within a team when working in small groups and maybe even in pairs. However, if the team grows or we have a high turnover of personnel, that might call for more documentation. If possible, processes should be automated, in which case they are also documented.

In general, agile projects do not have the same specialization in roles as on traditional projects. Everyone participates in all aspects of the project and should be able to carry out everything—at least in theory. This means that all developers should have sufficient knowledge about SCM to be able to carry out SCM-related activities by themselves. There will, for instance, not be a dedicated SCM-person to do daily or weekly builds or releases on an agile project. However, to do the design of the SCM-related work processes, even an agile team will need the help of an SCM expert who should work in close collaboration with the team such that individuals and interaction are valued over processes and tools (Agile Manifesto, 2001).

SCM Tools

In general, SCM is very process centric and could, in theory, be carried out by following these processes manually. However, agile methods try to automate the most frequently used processes and have tools take care of them (e.g., repository tools, build tools, automated merge tools, etc). Fortunately, the requirements that agile methods have to SCM tooling are not very demanding and can, more or less, easily be satisfied by most tools. For this reason, we do not want to give any tool recommendations or discuss specific tools, but rather focus on the general requirements and a couple of things to look out for. Furthermore, most often, you just use the tool that is given or the selection is based on political issues.

Using parallel work, we would need a tool that works without locking and thus has powerful merge capabilities to get as painless an integration as possible. Using test-driven development, we need to build very often so a fast build tool is very helpful—and preferably it will be flexible such that we can sometimes choose to trade speed for consistency. Working always against baselines, it would be nice if the repository tool would automatically handle bound configurations (Asklund, Bendix, Christensen, & Magnusson, 1999) so we should not do that manually.

However, a couple of things should be taken into account about SCM tooling. Because of refactoring and the fact that the architecture is grown organically, there will be changes to the structure of the files in the repository. This means that if the tool does not support name space versioning (Milligan, 2003), we will have a harder time because we lose history information and have no support for merging differing structures. However, this can be handled manually and by not carrying out structural changes in parallel with other work. It is much more problematic to actually change your repository tool in the middle of a project. Often you can migrate the code and the versions but you lose the meta-data that is equally as valuable for your work as the actual code. Therefore, if possible, you should try to anticipate the possible success and growth of the project and make sure that the tool will scale to match future requirements.

FUTURE TRENDS

While resisting the temptation to predict the future, we can safely assume that the increased use and awareness of SCM in agile development will result in a body of best practices and increased interaction between agile and SCM activities. Furthermore, we expect to see progress in tool support, including better merge support and increased traceability to name a few.

Continuous integration is an activity that has already received much attention and is quite mature and well understood. Many other SCM-related activities require continuous integration and we expect to see them mature accordingly when that foundation is now set. This will result in new best practices and perhaps specific SCM-related sub-practices to make these best practices explicit. A first attempt to specify SCM sub-practices for an agile setting is presented in Asklund, Bendix, and Ekman. (2004) and we expect them to mature and more sub-practices to follow.

SCM tools provide invaluable support and we envision two future trends. There is a trend to integrate various SCM-related tools into suites that support the entire line of SCM activities. These tools can be configured to adhere to pretty much any desired development process. They may, however, be somewhat heavyweight for an agile setting and as a contrast, we see the use of more lightweight tools. Most SCM activities described in this chapter can be supported by simple merge tools with concurrency detection.

Parallel work with collective code ownership can benefit from improved merge support. Current merge tools often operate on plain text at the file level and could be improved by using more fine-grained merge control, perhaps even with syntactic and partially semantics aware merge. An alternative approach is to use very fine-grained merge combined with support for increased awareness to lower the risk of merge conflicts. The increased use of SCM will also require merge support for other artefacts than source files.

The use of SCM in agile development will enable better support for traceability and tracking of changes. A little extra effort can provide bi-directional traceability between requirements, defects, and implementation. However, more experience is needed to determine actual benefits in an agile context before one can motivate and justify this extra "overhead."

SCM is being used more and more in agile methods, despite not being mentioned explicitly. However, it is often carried out in the same way as in traditional projects, but can benefit from being adapted to this new setting. The practices presented in this chapter adapt SCM for agile methods but more widespread use will lead to even more tailored SCM. In particular, SCM practices will be further refined to fit an agile environment and probably lead to more agile SCM. Some of these ideas may indeed transfer to traditional projects, providing more lightweight SCM in that setting as well.

CONCLUSION

SCM provides valuable activities that enhance the QA for agile development. The main quality enhancement does not stem directly from SCM but indirectly by supporting other quality enhancing activities. Traceability is, for instance, crucial to evaluate any kind of quality work, and configuration audits verify that SCM and QA activities have been carried out.

We have shown how typical agile activities can be supported directly by SCM techniques while retaining their agile properties. For instance, continuous integration demands support from SCM tools and processes to succeed while build and release management can help to streamline the release process to enable frequent releases. SCM can thus be used to support and strengthen such developer-oriented activities.

SCM is traditionally very strong in aspects that deal with the relation to the customer. Agile methods can benefit from these activities as well. Configuration control allows precise tracking of progress and traceability for each change request. Lightweight SCM plans simplify coordination within a team and help in effective use of other SCM-related activities. These are areas that are often not mentioned explicitly in agile literature.

There is, in general, no conflict between agile methods and SCM—quite the contrary. Agile methods and SCM blend well together and enhance each other's strengths. Safe SCM with rigorous change management can indeed be carried out in an agile project and be tailored to agile requirements.

SCM tools provide help in automating many agile activities, but we must stress that what is important are the SCM processes and not so much a particular set of tools. There are also many agile activities that could be supported even better by enhanced tool support. For instance, current merge tools are often fairly poor at handling structural merges such as refactorings; often this results in loss of version history and traceability, and incomprehensible merge conflicts.

Many agile teams already benefit from SCM, but we believe that a more complete set of SCM activities can be offered to the agile community.

Tailored processes and tools will add even more value and may indeed result in SCM activities that are themselves agile, which may even have an impact on more traditional software development methods.

REFERENCES

Adams, R., Weinert, A., & Tichy, W. (1989). Software change dynamics or half of all ADA compilations are redundant. *Proceedings of the 2nd European Software Engineering Conference*,Coventry, UK.

Agile Manifesto (2001). *Manifesto for agile software development.* Retrieved June 1, 2006, from http://agilemanifesto.org/

Aiello, B. (2003). Behaviorally speaking: Continuous integration: Managing chaos for quality! *CM Journal*, September.

Angstadt, B. (2000). SCM: More than support and control. *Crosstalk: The Journal of Defence Software Engineering*, March.

Appleton, B., & Cowham, R. (2004b). Release management: Making it lean and agile. *CM Journal*, August.

Appleton, B., Berczuk, S., & Konieczka, S. (2003a). Continuous integration: Just another buzz word? *CM Journal*, September.

Appleton, B., Berczuk, S., & Konieczka, S. (2003b). Codeline merging and locking: Continuous updates and two-phased commits. *CM Journal*, November.

Appleton, B., Berczuk, S., & Konieczka, S. (2004a). Continuous staging: Scaling continuous integration to multiple component teams. *CM Journal*, March.

Appleton, B., Berczuk, S., & Cowham, R. (2005). Branching and merging: An agile perspective. *CM Journal*, July.

Asklund, U., Bendix L., Christensen H. B., & Magnusson, B. (1999, September 5-7). The unified extensional versioning model. *Proceedings of the 9th International Symposium on System Configuration Management*, Toulouse, France.

Asklund, U., Bendix, L., & Ekman, T. (2004, August 17-19). Software configuration management practices for extreme programming teams. *Proceedings of the 11th Nordic Workshop on Programming and Software Development Tools and Techniques*, Turku, Finland.

Babich, W. A. (1986). *Software configuration management: Coordination for team productivity.* Addison-Wesley.

Beck, K. (1999a). Embracing change with extreme programming. *IEEE Computer*, *32*(10), 70-77.

Beck, K. (1999b). *Extreme programming explained: Embrace change.* Addison-Wesley.

Bellagio, D. E., & Milligan, T. J. (2005). *Software configuration management strategies and IBM Rational ClearCase.* IBM Press.

Bendix, L., & Vinter, O. (2001, November 19-23). Configuration management from a developer's perspective. *Proceedings of the EuroSTAR 2001 Conference*, Stockholm, Sweden.

Bendix, L., & Hedin, G. (2002). Summary of the subworkshop on extreme programming. *Nordic Journal of Computing*, *9*(3), 261-266.

Berczuk, S., & Appleton, S. (2003). *Software configuration management patterns: Effective teamwork, Practical Integration.* Addison-Wesley.

Berlack, H. R. (1992). *Software configuration management.* John Wiley & Sons.

Bohner, S. A., & Arnold, R. S. (1996). *Software change impact analysis.* IEEE Computer Society Press.

Buckley, F. J. (1993). *Implementing configuration management: Hardware, software, and firmware.* IEEE Computer Society Press.

Chromatic. (2003). *Chromatic: Extreme programming pocket guide.* O'Reilly & Associates.

Crnkovic, I., Asklund, U., & Persson Dahlqvist, A. (2003). *Implementing and integrating product data management and software configuration management.* Artech House.

Daniels, M. A. (1985). *Principles of configuration management.* Advanced Applications Consultants, Inc.

Dig, D., Nguyen, T. N., & Johnson, R. (2006). *Refactoring-aware software configuration management* (Tech. Rep. UIUCDCS-R-2006-2710). Department of Computer Science, University of Illinois at Urbana-Champaign.

Ekman, T., & Asklund, U. (2004). *Refactoring-aware versioning in eclipse.* Electronic Notes in Theoretical Computer Science, 107, 57-69.

Farah, J. (2004). Making incremental integration work for you. *CM Journal*, November.

Feiler, P. H. (1991). *Configuration management models in commercial environments* (Tech. Rep. CMU/SEI-91-TR-7). Carnegie-Mellon University/ Software Engineering Institute.

Feldman, S. I. (1979). Make—A program for maintaining computer programs. *Software—Practice and Experience, 9*(3), 255-265.

Fowler, M., & Foemmel, M. (2006). *Continuous integration.* Retrieved June 1, 2006, from http://www.martinfowler.com/articles/continuousIntegration.html

Hass, A. M. (2003). *Configuration management principles and practice.* Addison-Wesley.

Koskela, J. (2003). *Software configuration management in agile methods.* VTT publications: 514, VTT Tietopalvelu.

Leon, A. (2005). *Software configuration management handbook.* Artech House.

Mikkelsen, T., & Pherigo, S. (1997). *Practical software configuration management: The Late-night developer's handbook.* Prentice Hall.

Milligan, T. (2003). *Better software configuration management means better business: the seven keys to improving business value.* IBM Rational white paper.

Moreira, M. (2004). Approaching continuous integration. *CM Journal*, November.

Sayko, M. (2004). The role of incremental integration in a parallel development environment. *CM Journal*, November.

Schwanke, R. W., & Kaiser, G. E. (1988, January 27-29). Living with inconsistency in large systems. *Proceedings of the International Workshop on Software Version and Configuration Control*, Grassau, Germany.

Chapter VIII
Improving Quality by Exploiting Human Dynamics in Agile Methods

Panagiotis Sfetsos
Alexander Technological Educational Institution of Thessaloniki, Greece

Ioannis Stamelos
Aristotle University, Greece

ABSTRACT

Theory and experience have shown that human factors are critical for the success of software engineering practices. Agile methods are even more sensitive in such factors because they rely heavily on personal efforts with limited guidance from process manuals, allowing freedom in choosing solutions, inter-personal communications, etc. This fact raises important issues for the management of software engineers that are expected to apply agile methods effectively. One such issue at the agile organization executive level is human resource management, which should take into account agile development peculiarities, work competencies needed, agile workforce planning, etc. Another issue at the micro-management level is agile workforce management within the development process (e.g., team planning for a specific task or project) where individual human features will undoubtedly affect delivered quality and ultimately the task/project degree of success. This chapter deals with one problem at each level of management in an agile company applying extreme programming, one of the most diffused agile methods. In particular, the first part of the chapter proposes and discusses a model for personnel management based on the well known People-CMM[1] assessment and improvement model, while the second one proposes a model that exploits developer personalities and temperaments to effectively allocate and rotate developers in pairs for pair programming.

INTRODUCTION

Software engineering practices extensively involve humans under different roles (managers, analysts, designers, developers, testers, quality assurance experts, etc.) (Pfleeger, 2001; Sommerville, 2004). Software activities are still mostly based on individuals' knowledge and skills. On the other hand, in theory, agile methods put particular emphasis on people and their interactions. Agile organizations are expected to value individuals and interactions over processes and tools (Beck, 2000), but this fundamental consideration is often ignored and underestimated. Employment of people in agile projects presents both challenges and opportunities for managers. They should avoid pitfalls in managing agile software engineers such as assigning a developer to the wrong task, and they should exploit human competencies to assure high productivity and quality. As a consequence, people management is of paramount importance for agile organizations' success.

Often large organizations, applying both agile and traditional methodologies, have to integrate new processes with existing ones. These companies face cultural problems highlighted by differences between agile and traditional teams, and problems caused by distribution of work across multiple teams in large and complex projects (Cockburn, 2002; Highsmith, 2000; Lindval et al., 2004). On the other hand, small organizations are more dependent on skilled and experienced developers and are often facing problems related to human issues such as unpleasant conditions or relations among staff (Sfetsos, Angelis, & Stamelos et al., 2006a).

Regardless of its size, any organization applying agile methods must develop its own assessment and improvement processes for two reasons:

- To assure personnel quality at the corporate level, for example, to address workforce-related problems such as bad staffing, inadequate training, bad competency, and performance management, to mention some of the most important.
- To assure and exploit personnel qualities at the project/team level, for example, to identify early and understand the effects of its developer characteristics (skills, personalities, temperaments), effectively combining them to address problems quickly and improve communication and collaboration.

The rest of this chapter is organized in two separate sections. The first section deals with human resource management at the corporate level. It focuses on extreme programming (XP), which is analyzed from the perspective of the people capability maturity model (P-CMM), a five-level model that prescribes a framework for managing the development of people involved in software development processes. An analysis is provided showing that an XP organization, starting typically from the Managed Level (Level 2), would potentially successfully address most of the P-CMM Level 2 and 3 practices, and can reach Level 4 and 5 by applying quantitative measurements for improving performance of the competency-based processes. Eventually, an adaptive P-CMM assessment and improvement process model is proposed, which can be used by any XP organization for the successful implementation of its workforce management.

The second section provides a concrete example of how to assure quality at the project/team level; a pair formation and allocation model is built based on developer personalities and temperaments. A thorough and systematic analysis of human dynamics in pair programming, the most popular of XP practices, is provided aiming at the improvement of quality. First, the salient characteristics of the different personalities and temperaments on communication, knowledge management, and decision making in pair programming are analyzed. The results of a study investigating the impact of developer personalities and temperaments on communication and collaboration-viability in pair programming, using the Keirsey Temperament Sorter (KTS) (Keirsey & Bates, 1984), are reported. Next, an adaptive pair formation/rotation process model for the identification, interpretation, and the effective combination of developer variations to improve pair effectiveness is described.

ASSURING PERSONNEL QUALITY AT THE CORPORATE LEVEL: PEOPLE CAPABILITY MATURITY MODEL AND EXTREME PRAMMING

Evaluation and assessment (E&A) are critical activities in software engineering both for products and processes (Pfleeger, 2001). E&A models are also critical for organizations and people. They typically provide E&A structured in the form of levels; organizations achieve higher levels when they manage to design and effectively implement sets of processes and practices that are advanced with respect to those of the previous level. One widely accepted and used E&A model is CMM (Paulk, Curtis, Chrissis, & Weber, 1993), and its newest version CMM-I (Chrissis, Konrad, & Shrum, 2003).

As was discussed previously in Introduction, people, people quality, and people management are essential for agile companies. As a consequence, E&A people management models may help agile companies improve their people management processes and policies, assuring agile personnel quality. However, no such models capable to produce agile organization assessment have been proposed up to now. In the next, one such model, based on CMM people counterpart, namely People CMM, is outlined.

People CMM, first published in 1995 and revised 2001 (version 2) (Curtis, Hefley, & Miller, 1995, 2001), is a five-level model that focuses on

Table 1. Process areas of the People CMM: Version 2

Maturity Level	Focus	Key Process Areas
5 Optimizing	Continuously improve and align personal, workgroup, and organizational capability.	• Continuous workforce innovation. • Organizational performance alignment. • Continuous capability improvement.
4 Predictable	Empower and integrate workforce competencies and manage performance quantitatively.	• Mentoring. • Organizational capability management. • Quantitative performance management. • Competency-based assets. • Empowered workgroups. • Competency integration.
3 Defined	Develop workforce competencies and workgroups, and align with business strategy and objectives.	• Participatory culture. • Workgroup development. • Competency-based practices. • Career development. • Competency development. • Workforce planning. • Competency analysis.
2 Managed	Managers take responsibility for managing and developing their people.	• Compensation • Training and development. • Performance management. • Work environment. Communication and coordination staffing
1 Initial	Workforce practices applied inconsistently.	(no KPAs at this level)

continuously improving the management and development of the human assets of a software systems organization. People CMM, like most other capability maturity models, is a staged model for organizational change consisting of five maturity levels through which an organization's workforce practices and processes evolve. Each maturity level, representing a higher level of organizational capability, is composed of several *key process areas* (KPAs) that identify clusters of related workforce practices (see Table 1).

At the initial maturity level (Level 1), workforce practices are performed inconsistently and frequently fail to achieve their intended purpose. Managers usually rely on their intuition for managing their people. To achieve the managed maturity level (Level 2), the organization implements the discipline of performing basic workforce practices. While maturing to the defined level (Level 3), these practices are tailored to enhance the particular knowledge, skills, and work methods that best support the organization's business. To achieve the predictable maturity level (Level 4), the organization develops competency-based, high-performance workgroups, and empirically evaluates how effectively its workforce practices meet objectives. To achieve the optimizing maturity level (Level 5), the organization looks continually for innovative ways to improve its workforce capability and to support the workforce in their pursuit of professional excellence.

Practices of a key process area must be performed collectively achieving a set of goals considered important for enhancing workforce capability. People CMM can be applied by an organization in two ways: as a guide for implementing improvement activities and as a standard for assessing workforce practices. As a guide, the model helps organizations in selecting high-priority improvement actions, while as an assessment tool describes how to assess workforce capability. Due to limited space, the reader should consult (Curtis et al., 1995, 2001) for details about the key process areas. In the rest of this section we will examine XP from the People CMM perspective, presenting a brief summary of the KPAs effect in each maturity level and analyzing only those KPAs we consider successfully addressed by XP practices and values.

THE INITIAL LEVEL (MATURITY LEVEL 1)

At the initial level, work force practices are often ad hoc and inconsistent and frequently fail to achieve their intended purpose. This means that in some areas, the organization has not defined workforce practices, and in other areas, it has not trained responsible individuals to perform the practices that are established. Managers usually find it difficult to retain talented individuals and rely on their intuition for managing their people. Turnover is high so the level of knowledge and skills available in the organization does not grow over time because of the need to replace experienced and knowledgeable individuals who have left the organization.

XP is a high-disciplined methodology, thus organizations applying XP tend to retain skilled people, develop workforce practices, and train responsible individuals to perform highly cooperative best practices. Most of XP practices, especially pair programming, encourage the tacit transmission of knowledge and promote continuous training. Managers and coaches in XP organizations are well prepared to perform their workforce responsibilities. We consider that most XP organizations bypass the initial level.

KEY PROCESS AREAS AT THE MANAGED LEVEL (MATURITY LEVEL 2)

The key process areas at managed level focus on establishing basic workforce practices and eliminating problems that hinder work performance.

At the managed level, an organization's attention focuses on unit-level issues. An organization's capability for performing work is best characterised by the capability of units to meet their commitments. This capability is achieved by ensuring that people have the skills needed to perform their assigned work and by implementing the defined actions needed to improve performance. *Staffing* is designed to establish basic practices by which talented people are recruited, selected among job candidates, and assigned to tasks/projects within the organization. Knowledge intensive XP organizations setting skill requirements at a higher level must coordinate their staff selection activities to attract developers capable to implement demanding XP practices, such as pair programming, test driven development, etc. The purpose of *communication* is to establish a social environment that supports effective interaction and to ensure that the workforce has the skills to share information and coordinate their activities efficiently. In the XP process, communication is the most significant of the four prized values, starting from the early phase of the development process (planning game) and being implemented in most of the other practices (i.e., pair programming, testing, etc.). The purpose of *work environment* is to establish and maintain physical working conditions that allow individuals to perform their tasks efficiently without distractions. For XP teams (usually small, 2-12 persons), one large room with small cubbies at the side is used. All team members (programmers, coach, customer, etc.) work together in this room. *Performance management* is designed to establish objective criteria against which unit and individual performance can be measured, and to enhance performance and feedback continuously. Skills obtained by the successful implementation of XP practices are capable to boost performance. XP addresses with success requirement changes through user stories in planning game, continuous integrations, and small releases. Pair programming with continuous code reviews, faster code production, and learning of both programming techniques and problem domain increases performance. Testing, minimizing defect rates, and on-site customer providing feedback often and early are also significant factors affecting positively performance. The same effect is obtained with simple design, common code ownership, and metaphor. The purpose of *training and development* is to ensure that all individuals have the skills required to perform their assignments. XP addresses successfully training needs by rotating developers in pair programming and by involving them in significant practices such as planning game, testing, refactoring, and metaphor. *Compensation* is designed to provide all individuals with payment and benefits based on their contribution and value to the organization. Apart from compensation, 40-hours a week is a practice benefiting both developers and organization. Consequently, we consider that XP organisations would address P-CMM Level 2 KPAs without problems.

KEY PROCESS AREAS AT THE DEFINED LEVEL (MATURITY LEVEL 3)

In order to mature into the defined level, basic workforce practices that have been established for units (managed level) are tailored to enhance the particular knowledge, skills, and work methods that best support the organization's business. At this level, organization addresses organizational issues, developing a culture of professionalism based on well-understood workforce competencies. *Competency analysis, competency development, and competency-based practices* are designed to identify, develop, and use the knowledge, skills, and process abilities required by workforce to perform the organization's business activities, respectively. *Career development* is designed to ensure that individuals are provided opportunities to develop workforce competencies enabling them to achieve career objectives. *Workgroup develop-*

ment on the other hand strives to organize work around competency-based process abilities. All the previously mentioned process areas contribute in creating a *participatory culture*, which gives the workforce full capability for making decisions that affect the performance of business activities. They also assist in *workforce planning*, which refers to coordination of workforce activities with current and future business needs. An XP organization can enhance workforce competencies by:

- Providing opportunities for individuals to identify, develop, and use their skills and knowledge involving them in the implementation of the XP practices, and by
- Using the skills and knowledge of its workforce as resources for developing the workforce competencies of others (e.g., through pair programming).

XP teams, amalgamating technical and business people with divergent backgrounds and skills, keep the most significant role in identification, development, and use of competency practices. *Competency practices* starts with pair programming that helps managers and developers to identify, develop, and use available knowledge and skills. Technical competencies related to methodologies, project-based knowledge, and tool usage are improved by planning game, pair programming, test-driven development, refactoring, simple design, and common code ownership. Knowledge and skills, obtained by gradual *training* and successful projects, enhance organization's knowledge repository. The XP process establishes a high *participatory culture* (pair programming and other practices), spreading the flow of information within the organization, and incorporating the knowledge of developers into decision-making activities, providing them with the opportunity to achieve *career objectives.* Iterative and incremental development with small releases assist in *work-force planning,* which refers to coordination and synchronization of workforce activities with current and future business needs. Consequently, we consider that XP organizations are well prepared to successfully address most of the P-CMM Level 3 KPAs.

KEY PROCESS AREAS AT THE PREDICTABLE LEVEL (MATURITY LEVEL 4)

In maturing to the predictable level, the organizational framework of workforce competencies that has been established in the defined level is both managed and exploited. The organization has the capability to predict its performance and capacity for work even when business objectives are changed through a culture of measurement and exploitation of shared experiences. The key processes introduced in this level help organizations quantify the workforce capabilities and the competency-based processes it uses in performing its assignments. *Competency integration* is designed to improve the efficiency and agility of interdependent work by integrating the process abilities of different workforce competencies. The purpose of *empowered workgroups* is the creation of workgroups with the responsibility and authority to determine how to conduct their business activities more effectively. *Competency-based assets* is designed to capture the knowledge, experience, and artefacts developed while performing competency-based processes for enhancing capability and performance. *Quantitative performance management* is designed to predict and manage the capability of competency-based processes for achieving measurable performance objectives. *Organizational capability management* is designed to quantify and manage the capability of the workforce and of the critical competency-based processes they perform. *Mentoring* is designed to transfer the lessons obtained through experience into a work-force competency to improve the capability of other individuals or workgroups.

XP is a team-based process helping workgroups to develop more cohesion, capability, and responsibility. Team-based practices, competency practices, training, and mentoring are the key process areas most benefiting from pairing. Mentoring in pair programming is a never-ending process, transferring inter-personal knowledge in an informal way (Williams & Kessler, 2002; Williams, Kessler, Cunningham, & Jefferies, 2000). Recent research studies have shown that the assimilation time came down from 28 days to 13 days, the mentoring time was reduced from 32% to 25%, and the training effort was cut down by half (Williams et al., 2002). XP process requires that developers implement best practices in extreme levels using proven *competency-based activities* in their assignments. Managers trust the results that developers produce and the XP organization preserves successful results in its repository and exploits them as organizational assets. Organizational assets can be used effectively again and again as corporate standards, increasing productivity and spreading learning rapidly through the organization. Managers trusting team competencies *empower teams* by transferring to them responsibility and authority for performing committed work. Developers define milestones for coordination, integrating their competency-based activities into a single process. This process, constituted from different workforce competencies, should be institutionalized by organization, which begins to manage its capability quantitatively. The performance of each unit and team should be measured enabling organizations performance to become more predictable. The integration of the people processes with business processes and measuring of the co-relations between the two will help an XP organization to mature up to this level.

KEY PROCESS AREAS AT THE OPTIMIZING LEVEL (MATURITY LEVEL 5)

The process areas at the optimizing level focus on continuous improvement of workforce capability and practices. These practices cover issues that address continuous improvement of methods for developing competency at both the organizational and the individual level. The organization uses the results of the quantitative management activities established at level 4 to guide improvements at this level. *Continuous capability improvement* provides a foundation for individuals and workgroups to continuously improve their capability for performing competency-based processes. *Organizational performance alignment* enhances the alignment of performance results across individuals, workgroups, and units with organizational performance and business objectives. *Continuous workforce innovation* is designed to identify and evaluate improved or innovative workforce practices and technologies, and implement the most promising ones throughout the organization.

XP practices, especially pair programming with pair rotation, help increasing the knowledge level of the individuals and subsequently of the team. As mentioned in level 4, this knowledge enriches organization's knowledge repository providing both individuals and workgroups the ability to *continuously improve* their *capabilities.* This improvement occurs through incremental advances from the implementation of the XP practices. The results from measurements at level 4 and the *culture of improvements* established by the continuous implementation of the XP practices can help the XP organization to mature up to this level.

Figure 1. An adaptive people CMM assessment process model for assessing XP- organizations

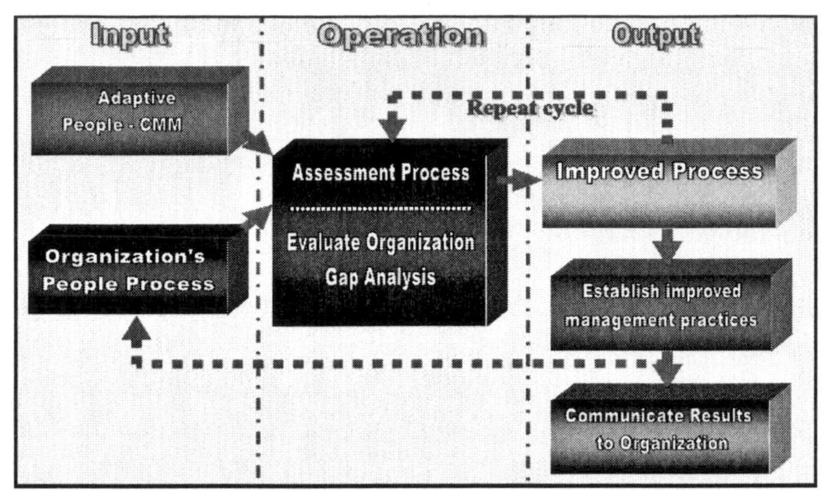

AN ADAPTIVE PEOPLE CMM ASSESSMENT PROCESS MODEL TO ASSESS XP ORGANIZATIONS

The process model we suggest is an adaptive people CMM assessment process model in the sense that the XP organization assesses itself against the process areas defined in each maturity level (focusing mostly on those previously discussed), and decides what course of action to take and how to address the improvement areas. The model (see Figure 1) is divided into three stages:

- **Input**, where the people process currently used by the XP organization and the adaptive people CMM framework are entered into the process.
- **Operation**, where the assessment process takes place.
- **Output**, where the results of the assessment process, in the form of a new improved process, are adopted by the people process management task and are communicated to the organization.

The main assessment process starts with a *gap analysis* (Curtis et. al., 1995, 2001), where organization's workforce activities are examined against people CMM to identify gaps or shortcomings. This kind of analysis helps the organization to measure progress. Gap analysis can be conducted as a guided workshop session led by a qualified assessor or facilitator. Typical steps are:

1. An assessor or a small team of assessors consisting of managers and developers is selected and trained in the people CMM. After a short presentation describing the People CMM and the purpose of the survey, the program manager or facilitator assigns a specific process area and the proper evaluation questionnaire to assessors.
2. Each assessor scores and comments on the process areas individually in the questionnaire, evaluating the organization against each criteria item, and determining how well the organization satisfies the described practices. Questionnaires can be filled in a group session.
3. Program manager or facilitator picks up questionnaires, elaborates scores and comments analyzing responses, and prioritizes results for discussion.
4. Program manager or facilitator convokes a consensus meeting focusing on key areas with low scores (i.e., areas needing improve-

ment). Meeting leads to agreement on key improvement areas based on overall assessment and comes to a consensus on prioritized inputs.

5. Summary reports are written, describing the results for both the developer and the manager questionnaires. These reports provide team members with information about the consistency with which workforce practices are performed and about the major issues related to them. Reports provides both summary statistical data and written comments related to questions.

The assessment results are firstly incorporated into the work-force practices and secondly the improved workforce practices are established and communicated to the organization. Analytically:

- Responses are analyzed and a summary presentation is delivered to the organization.
- The recommended action plans and detailed improvement activities are prioritized and incorporated into the workforce management task cycle plan to address identified areas for improvement. These steps in the assessment must be repeated in short period times (i.e., every year) to keep the assessment up to date, to evaluate the progress of previously deployed assessment processes, and to use the results to feed the yearly planning cycle.

After the application of the improved process, the next step is to move into the establishing phase where a program of continuous workforce development is established. In this phase, a program of workforce development is integrated with corporate process improvement, linking together improved workforce practices with organization's workforce process. Improved workforce practices are continuously used incorporating a culture of excellence. The step is to move into the communication phase where strengths, shortcomings, changes in organizational structure or processes, action plans, and detailed actions that must be taken to improve practices are communicated to the entire organization.

EXPLOITING PERSONNEL QUALITIES AT THE PROJECT/TEAM LEVEL: ASSESSING AND IMPROVING PAIR PROGRAMMING EFFECTIVENESS BASED ON DEVELOPER PERSONALITIES

As discussed in the Introduction, one of agile organizations major concerns must be careful personnel management at the project/team level. Apart from correct handling of individual technical skills, how could developer personality and temperament types be used to obtain improved performance and ultimately increased software quality levels? This section exemplifies such personnel treatment by providing a model for pair formation and allocation in pair programming.

Human Issues in Pair Programming

Extreme programming bases its software development process on a bunch of intensely social and collaborative activities and practices (Beck, 2000). The intent of these practices is to capitalize on developer's unique skills, experiences, idiosyncrasies, and personalities, considering them as the first-order impact on project success. Pair programming, a popular practice not only in XP, is a disciplined practice in which the overall development activity is a joint effort, a function of how people communicate, interact, and collaborate to produce results.

In the past few years, pair programming has received increased interest not only as a best practice in extreme programming, but also as a standalone programming style. It is an intensively social and collaborative activity practiced by two developers working together at one machine (Beck, 2000). One of the developers is the driver—creating artefacts (e.g., code, designs), and the other is the navigator—peer reviewing the driver's output and thinking of alternatives. Developers must periodically switch roles and partners so that the overall development activity is a joint effort. Creativity becomes a function of how developers communicate, interact, and collaborate to produce results (Beck, 2000). When working in pairs, their personal preferences, traits, and characteristics have a strong influence on their decisions and actions.

Up to now, organizations and managers have faced pair programming as a rough technical process (Sfetsos et al., 2006a, Sfetsos, Stamelos, Angelis, & Deligiannis, 2006b). But as in any software process, there exist human factors that can not be easily identified and understood well enough to be controlled, predicted, or manipulated. On the other hand, performance and effectiveness problems always exist and must be addressed successfully. Such problems are not addressable through the known improvement approaches, as most of them focus on processes or technology, not on people. The primary role of people has been largely ignored up to now and no efforts have been devoted to increase developers' communication, collaboration, and ultimately effectiveness or to address pair problems and failures. Beck states that management has much to gain from psychology to understand where and why slowdowns occur (Beck, 2000). Cockburn claims that only developers with different personalities and with the same experience, if effectively combined, can minimize communication gaps (Cockburn, 2002). This means that management must utilize processes, which first identify and understand developers' personalities and then effectively combine their potential strengths, fostering communication and collaboration. However, one critical question that still remains to be answered is which personality types should be combined in pair formations and rotations?

In the rest of the chapter, we will try to answer this research question and we will propose an adaptive pair formation/rotation process model for the identification, interpretation, and the effective combination of developer variations. We base our approach on Cockburns' team ecosystems as described in his *Teams as Ecosystems* (Cockburn, 2002), on the findings of two field studies, the first at the North Carolina State University (Katira et al., 2004) and the second at 20 software development teams in Hong Kong (Gorla & Lam, 2004), on the findings of a survey of 15 agile companies (Sfetsos et al., 2006a), and on the results of a controlled experiment we conducted (Sfetsos et al., 2006b).

We consider pairs as *adaptive ecosystems* in which physical structures, roles, and developer personalities all exert forces on each other. They are adaptive because developers through pair rotations, can create, learn, and respond to change. In these ecosystems, the overall development activity becomes a joint effort, a function of how paired developers communicate, interact, and collaborate to produce results. However, different personalities express different natural preferences on communication, information, and knowledge handling and sharing, decision-making, and problem solving (Cockburn, 2002; Highsmith, 2002). Personalities are not right or wrong, they just are, and can be more or less effective, more or less appropriate for different roles and tasks. They can be turned into powerful tools instead of dividing obstacles, contributing to success if effectively combined (Ferdinandi, 1998). By understanding developer variations and knowing what motivates them, we can facilitate the pair formation and pair rotation process, allowing individuals to work in areas in which they are strong. Laplante and Neil claim that: "Having understood people motivations, it becomes easier to seek win-win solutions or avoid causing problems" (Laplante & Neil, 2006).

Pair Programming Roles and Actions

Paired developers must succeed in many formal or informal assigned roles, either pair[2] or functional[3], such as the role of a leader, mentor, coordinator, facilitator, innovator, analyser, tester, decision maker, negotiator, and that of a peer reviewer, to mention some of the most significant. To accomplish all these different roles, developers must deploy a broad set of interpersonal skills, which complement each other, ensuring effective pair interrelationship and cohesion. Literature does not provide guidelines for the optimal distribution of roles and tasks among the paired developers. However, managers should assign roles and tasks according to the strong points of developer personalities effectively combining their talents and strengths in pair rotations.

Communication and Collaboration

Communication is one of the four prized values in XP, but its impact on pair performance and effectiveness has not been empirically investigated. In particular, pair programming assumes that developers with frequent, easy, face-to-face communication will find it easier to develop software, get quick feedback, and make immediate corrections in their development course. But as software grows and pairs rotate, communication paths spread and grow, thus increasing the effort for successful communication. Therefore, collaboration and personal contact among developers must be further improved, eliminating problems and smoothening possible differences and conflicts. Developer communication, as people communication in general, is never perfect and complete depending on developers' personality preferences.

Collaboration is defined as an act of shared creation. It differs from communication in the sense that it involves joint and active participation in all paired activities, especially in the creation of working software, in decision-making, and in knowledge management (Highsmith, 2002). In pair programming, many important decisions, which must be made quickly and well are often left to developers. Decisions are made, but the question is what criteria are used and what is the scope of the decisions. Managers must facilitate pair decision-making, taking into account developer personality preferences and motivations, in addition to the level of knowledge and information possessed by the pair, linking successful decisions to good performance and effectiveness. The same holds for transferring and sharing knowledge. During pair programming sessions, explicit and tacit knowledge are transferred and shared between developers. Tacit knowledge is managed first through face-to-face communication and subsequently through developer rotation, simple workable code, and extensive unit tests.

Identifying and Understanding Personalities and Temperaments

Two widely used tools to assist in the identification of personality and temperament types are the Myers-Briggs Type Indicator (MBTI[4]) (Myers, 1975) and the Keirsey Temperament Sorter (KTS) (Keirsey et al., 1984). The MBTI, a 94-item questionnaire, focuses on four areas of opposite behavior preferences forming 16 different personality types. It is used to quickly identify where people get their energy, how they gather information, how they make decisions, and which work style they prefer. The four pairs of preferences are *Extraverting* (E) and *Introverting* (I), *Sensing* (S) and *iNtuiting* (N), *Thinking* (T) and *Feeling* (F), and *Judging* (J) and *Perceiving* (P). The KTS, a 70-item questionnaire, classifies the 16 personality types into four temperament types: *Artisan* (SP), *Guardian* (SJ), *Idealist* (NF), and *Rational* (NT). We used the hardcopy of the Keirsey Temperament Sorter[5] to identify and interpret the personality inventories of the participants in one experiment with pair programming. In Table 2, we summarise the salient characteristics of each

Table 2. The salient characteristics of personality types with respect to pair programming

Personality Type	Salient Characteristics	Suggested use in Pair Programming
Extroverts	• Get energy from the outside world, experiences, and interactions. • Talk easily.	• Suitable for interactions with users and management. • May be good drivers.
Introverts	• Get energy from within themselves, from internal thoughts, feelings, and reflections. • Prefer finished ideas, prefer to read and think about something before start talking. • Prefer to be silent.	• Might not be suitable for pair programming. • Must be handled with care in meetings. • May become navigators.
Sensors	• Gather information linearly through senses. • Observe what is happening around. • Recognize the practical realities of a situation. • Take things literally and sequentially. • Concentrate on details. • Prefer tangible results clearly described.	• Probably the most capable programmers.
Intuitives	• Gather information more abstractly • See the big picture of a situation or problem. • Focus on relationships and connections between facts. • Good at seeing new possibilities and different ways of doing things.	• Probably the most capable system and application analysts.
Thinkers	• Make objective decisions. • Are logical, critical, and orderly. • Prefer to work with facts. • Examine carefully cause and effect of a choice or action. • Can apply problem-solving abilities.	• Suitable for making pair decisions. • Suitable for problem-solving situations.
Feelers	• Make subjective decisions. • Are driven by personal values. • Likes to understand, appreciate, and support others. • Are more people-oriented.	• Are good pair and team-builders. • Are good in relations with other pairs.
Judgers	• Live in an orderly and planned way, with detailed schedules. • Prefer things decided and concluded. • Prefer to avoid last-minute stresses.	• May be good navigators. • Generally combines well with a perceiver.
Perceivers	• Live in a flexible, spontaneous way. • Rely on experience. • Leave open issues. • Explore all possibilities. • Find difficulty with decision-making. • Often relies on last minute work.	• May be good drivers. • Generally combines well with a Judger.

personality type and our suggestions for exploiting them in pair programming.

In Table 3, we summarize the temperaments salient characteristics and our suggestions for their use in pair programming.

It is good to have variety of pairs—extroverts and introverts, abstract and concrete thinkers, orderly and random approaches—with people who enjoy diving into details before deciding and others who decide quick and are guided by perception. Therefore, it is up to organizations and managers to effectively combine developer diversities in pair rotations, allowing individuals to work in roles and tasks in which they can actually succeed.

An Adaptive Pair Formation and Rotation Process Model

In a recent field research study (Sfetsos et al., 2005a), we found out that software companies applying pair programming experienced problems due to human factors. In interviews, developers pinpointed that the most important problem they

Table 3. The salient characteristics of temperament types with respect to pair programming

Temperament Type	Salient Characteristics	Suggested use in Pair Programming
Artisans (SP) *(Sensing-Perceiving)*	• Prefer concrete communications. • Prefer a cooperative path to goal accomplishment. • Possess a superior sense of timing. • Prefer practical solutions. • Are lateral thinkers.	• Good as start-up persons. • Effective brainstormers. • May be good in decision making. • May exhibit adaptability and be innovative.
Guardians (SJ) *(Sensing-Judging)*	• Prefer concrete communications. • Prefer more a utilitarian approach. • Are traditionalists and stabilizers. • Prefer rules, schedules, regulations, and hierarchy. • Prefer that things remain as are.	• May be good in estimations (e.g. from user stories). • May be good in resource management. • May be good in planning game, contracts. • Are considered very responsible, succeed in assigned tasks.
Idealists (NF) *(Intuitive-Feeling)*	• Prefer more abstract communications. • Prefer more a utilitarian approach. • Prefer to guide others. • Excellent communicators.	• Will contribute to pair spirit and morale. • Are good in personal relationships. • Are good in interaction with users and management. • May be forward and global thinkers.
Rationalists (NT) *(Intuitive-Thinking)*	• Prefer more abstract communications. • Prefer a cooperative path to goal accomplishment. • Are natural-born scientists, theorists, and innovators. • Possess highly valuing logic and reason. • Prefer competence and excellence.	• Are good in subtask identification. (e.g., in splitting user stories) • Are good in long-range plans (i.e., planning game) • Are good in analysis and design. • Are considered good in inventing and configuring.

are facing is the unpleasant relations with their pair-mates. Besides, managers stated that such problems can not be addressed easily because most improvement programs focus on processes or technology, not on people. However, in general, literature and published empirical studies on pair programming do not delve in issues concerning developers' personalities and temperaments and how they should be effectively combined, so as to match their potential roles and tasks. In another recent case study (Katira et al., 2004), it was observed that undergraduate students seem to work better with partners of different personality type. In order to obtain concrete evidence that supports or rejects the hypothesis that the combination of developers with different personalities and with the same experience can minimize communication and collaboration gaps, we conducted a formal controlled experiment with the participation of 84 undergraduate students. The objective of the experiment was to compare pairs comprised of mixed personalities with pairs of the same personalities, in terms of pair effectiveness (Sfetsos et al., 2006b). *Pair effectiveness* (similar to team effectiveness, Sundstrom, Meuse, & Futrell, 1990) was captured through: *pair performance*—measured by communication, velocity, productivity, and customer satisfaction (passed acceptance tests), and *pair viability*—measured by developers' satisfaction, knowledge acquisition, and participation (communication satisfaction ratio, nuisance ratio, and driver or navigator preference).

Considering the importance of communication in pair performance, we included the communication variable in the experiment variables system. The results of the experiment have shown that there is significant difference between the two groups, indicating better performance and viability for the pairs with mixed personalities.

Based on the findings of the three field studies, the results of the experiment and having the theory that considers pairs as adaptive ecosystems as framework, we propose an adaptive pair formation/rotation process model (see Figure 2). This model can help organizations and managers build high-performance pairs out of talented developers. It describes three main phases: the setup phase, the assessment phase, and the improvement phase. The setup phase includes the identification, understanding, and interpretation of the developer personalities—temperaments. The assessment phase includes a gap analysis and the construction or review of a set of guidelines and policies for pair formation/rotations. The improvement phase includes mini retrospectives (communication-collaboration reviews) for pair evaluation, and the establishment of the improved pair rotation process. In detail, the set of actions, which must be successively taken are:

1. Identify developer personalities and temperaments using the KTS or the MBTI tool, creating personality and temperament inventories.
2. Understand and interpret the impact of developer personalities and temperaments on communication and collaboration using existing personality and temperament inventories to find their strong and weak points.
3. Assess existing situation analytically:

- Perform gap analysis. First start noticing developer strengths, weaknesses, and oddities. Notice how some developers:
  - Fit their roles and task.
  - Exhibit a steady performance.

Figure 2. An adaptive pair formation/rotation process model

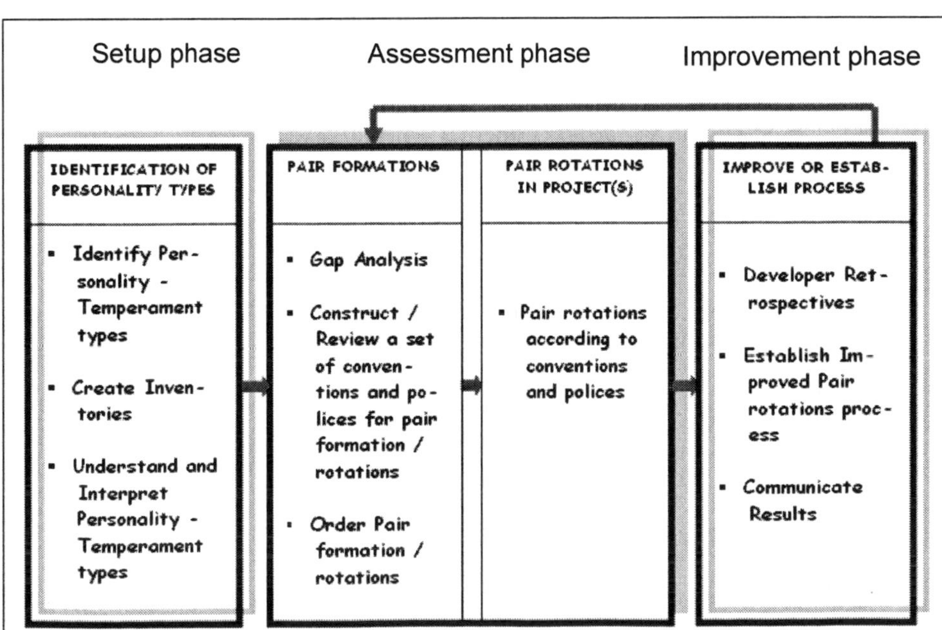

- o Take unnecessary risks, while others are conservative.
- o Construct a set of conventions and policies that might work well for them, suiting their strengths and weaknesses.
- o Order pair formations/rotations for pair programming projects, combining strengths to minimize weaknesses, assigning the roles and tasks to developers by their strong points of their personalities.

4. Monitor developer and pair performance in regular mini retrospectives (communication-collaboration reviews), helping developers learn about themselves and how they will effectively communicate and collaborate. Retrospectives for people reviews are used in ASD (Adaptive Software Development (Highsmith, 2000) and DSDM (Dynamic Systems Development Method) (Stapleton, 1997).
5. Establish improved pair formation/rotation process, communicate the results.

CONCLUSION

In the first part of this chapter, we analysed XP from the P-CMM perspective and proposed an adaptive P-CMM assessment and improvement process model for improving workforce quality in XP organizations, providing stepwise guidelines for its implementation. An agile organization's maturity from the P-CMM perspective derives from the repeatedly performed workforce practices, and the extent to which these practices have been integrated into the organizations' repository. The more mature an organization, the greater its capability for attracting, developing, and retaining skilled and competent employees it needs to execute its business. Agile methods, in particular extreme programming, through their repeatable practices lead to an improved workforce environment with learning, training, and mentoring opportunities, improving workforce competencies. We believe that organizations practicing XP should not have problems in addressing most of the P-CMM level 2 and 3 KPAs. XP organizations, starting usually from the managed level (level 2), have to make relatively limited adjustments in their workforce practices to manage other key process areas. Using measures on the performance of competency-based processes can mature an XP organization into level 4. The continuous improvement of competency-based processes, using the results of measurements, can mature an XP organization into level 5. We described an adaptive people CMM assessment process model for assessing XP organizations and stepwise guidelines for its implementation.

In the second part of the chapter, we focused on human factors in pair programming, the heart of the XP practices' implementation. Considering pairs as adaptive ecosystems, we investigated how developers with different personalities and temperaments communicate, interact, and collaborate to produce results. In particular, we established the impact of developers' natural preferences and traits on the assigned roles, communication, decision-making, and knowledge management. Based on the findings of three field studies, the results of an experiment, and using as framework the theoretical background of agile methods, we propose an adaptive pair formation/rotation process model, which identifies, interprets, and effectively combines developer variations. The proposed model can help organizations and managers improve pair effectiveness, by matching developers' personality and temperament types to their potential roles and tasks, effectively exploiting their differences in pair formations and rotations.

REFERENCES

Beck, K. (2000). *Extreme programming explained: Embrace change.* Reading, MA: Addison-Wesley.

Chrissis, M. B., Konrad, M., & Shrum, S. (2003). *CMMI: Guidelines for process integration and product improvement.* Boston: Addison-Wesley.

Cockburn, A. (2002). *Agile software development.* Boston: Addison-Wesley.

Curtis, B., Hefley, W. E., & Miller, S. (1995, September). *People capability maturity model* (CMU/SEI-95-MM-002 ADA300822). Pittsburgh, PA: Software Engineering Institute, Carnegie Mellon University.

Curtis, B., Hefley, W. E., & Miller, S. (2001, July). *People capability maturity model Version 2.0,* (CMU/SEI-2001-MM-01). Pittsburgh, PA: Software Engineering Institute, Carnegie Mellon University.

Ferdinandi, P. (1998, September/October). Facilitating collaboration. *IEEE Software*, 92-98.

Gorla, N., & Lam, Y. W. (2004, June). Who should work with whom? Building effective software project teams. *Communications of ACM, 47*(6), 79-82.

Highsmith, J. (2000). *Adaptive software development: A collaborative approach to managing complex systems.* New York: Dorset House.

Highsmith, J. (2002). *Agile software development ecosystems.* Boston: Addison Wesley.

Katira, N., Williams, L., Wiebe, E., Miller, C., Balik, S., & Gehringer, E. (2004). On understanding compatibility of student pair programmers. *SIGCSE'04* (pp. 3-7).

Keirscy, D., & Bates, M. (1984). *Please Understand Me*, Del Mar, California: Prometheus Book Company.

Laplante, P., & Neil, C. (2006). *Antipatterns. Identification, refactoring, and management.* Boca Raton, FL: Auerbach Publications.

Lindvall, M., Muthig, D., Dagnino, A., Wallin, C., Stupperich, M., Kiefer, D., May, J., & Kähkönen, T. (2004, December). Agile software development in large organizations. *Computer, IEEE, 37*(12), 26-24.

Myers, I. (1975). *Manual: The Myers-Briggs type indicator.* Palo Alto, CA: Consulting Psychologists Press.

Paulk, M. C., Curtis, B., Chrissis, M. B., & Weber, C. V. (1993). *Capability maturity model for software, Version 1.1.* Software Engineering Institute: Capability Maturity Modeling, 82.

Pfleeger, S. (2001). *Software engineering: Theory and practice* (2nd ed.). NJ: Prentice-Hall, Inc.

Sfetsos, P., Angelis, L., & Stamelos, I. (2006a, June). Investigating the extreme programming system—An empirical study. *Empirical Software Engineering, 11*(2), 269-301.

Sfetsos, P., Stamelos, I., Angelis, L., & Deligiannis, I. (2006b, June). Investigating the impact of personality types on communication and collaboration—Viability in pair programming—An empirical study. The *7th International Conference on eXtreme Programming and Agile Processes in Software Engineering* (XP2006), Finland.

Sommerville, I. (2004). *Software engineering* (7th ed.). Addison Wesley.

Stapleton, J. (1997). *DSDM, dynamic systems development method: The method in practice.* Harlow, UK: Addison-Wesley.

Sundstrom, E., De Meuse, K., & Futrell, D. (1990, February). Work teams. *American Psychologist, 45*, 120-133.

Williams, L., & Kessler, R. (2002). *Pair programming illuminated.* Boston: Addison-Wesley.

Williams, L., Kessler, R., Cunningham, W., & Jefferies, R. (2000, July/August). Strengthening the case for pair-programming. *IEEE Software, 17,* 19-25.

ENDNOTES

1. The people capability maturity model (P-CMM) was developed by the Software Engineering Institute (SEI) at Carnegie Mellon University (Curtis et al., 1995, 2001).
2. Roles that developers must undertake into pairs, usually informally assigned (e.g., leader, mentor).
3. Roles defined by the individual's technical skills and knowledge (e.g., tester).
4. Myers-Briggs type indicator and MBTI are registered trademarks of the Myers-Briggs type indicator trust.
5. See http://keirsey.com/cgi-bin/keirsey/kcs.cgi

Chapter IX
Teaching Agile Software Development Quality Assurance

Orit Hazzan
Technion – Israel Institute of Technology, Israel

Yael Dubinsky
Technion – Israel Institute of Technology, Israel

ABSTRACT

This chapter presents a teaching framework for agile quality—that is, the way quality issues are perceived in agile software development environments. The teaching framework consists of nine principles, the actual implementation of which is varied and should be adjusted for different specific teaching environments. This chapter outlines the principles and addresses their contribution to learners' understanding of agile quality. In addition, we highlight some of the differences between agile software development and plan-driven software development in general, and with respect to software quality in particular. This chapter provides a framework to be used by software engineering instructors who wish to base students learning on students' experiences of the different aspects involved in software development environments.

INTRODUCTION

Quality assurance (QA) is an integral and essential ingredient of any engineering process. Though there is a consensus among software practitioners about its importance, in traditional software development environments conflicts may still arise between software QA people and developers (Van Vliet, 2000, p. 125).

Agile software development methods emerged during the past decade as a response to the characteristics problems of software development processes. Since the agile methods introduced a different perspective on QA, we will call the agile approach toward quality issues *agile quality*—AQ, and will focus, in this chapter, on the teaching of AQ. By the term AQ, we refer to all the activities (e.g., testing, refactoring, requirement gathering) that deal with quality as they are manifested and applied in agile software development environments. It is important to emphasize that the term AQ does not imply that quality changes. To the contrary, the term AQ reflects the high standards that agile software methods set with respect to software quality.

Based on our extensive experience of teaching agile software development methods both in academia and in the software industry[1], we present a teaching framework for AQ. The teaching framework consists of nine principles, the actual implementation of which is varied and should be adjusted for different specific teaching environments (e.g., academia and industry to different sizes of groups). This chapter outlines the principles and addresses their contribution to learners' understanding of AQ.

In the next section, we highlight some of the differences between agile software development and plan-driven[2] software development in general, and with respect to software quality in particular. Then, we focus on the teaching of AQ. We start by explaining why quality should be taught and, based on this understanding, we present the teaching framework for AQ, which suggests an alternative approach for the teaching of AQ. Finally, we conclude.

Agile vs. Plan-Driven Software Development

In this section, we highlight some of the main differences between agile software development and traditional, plan-driven software development. Before we elaborate on these differences, we present our perspective within which we wish to analyze these differences.

Traditional software development processes mimic traditional industries by employing some kind of production chain. However, the failure of software projects teaches us that such models do not always work well for software development processes. In order to cope with problems that result from such practices, the notion of a production chain is eliminated in agile software development environments and is replaced by a more network-oriented development process (Beck, 2000). In practice, this means that in agile teams, the task at hand is *not* divided and allocated to several different teams according to their functional description (for example, designers, developers, and testers), each of which executes its part of the task. Rather, all software development activities are intertwined and there is no passing on of responsibility to the next stage in the production chain. Thus, all team members are equally responsible for the software quality. We suggest that this different concept of the development process results, among other factors, from the fact that software is an intangible product, and therefore it requires a different development process, as well as a different approach toward the concept of software quality, than do tangible products.

Agile Development Methods vs. Plan-Driven Development Methods

During the 1990s, the agile approach toward software development started emerging in response to the typical problems of the software industry. The approach is composed of several methods and it formalizes software development frameworks that aim to systematically overcome characteristic problems of software projects (Highsmith, 2002). Generally speaking, the agile approach reflects the notion that software development environments should support communication and information sharing, in addition to heavy testing, short releases, customer satisfaction, and sustainable work-pace for all individuals involved in the process. Table 1 presents the manifesto for agile software development (http://agilemanifesto.org/).

Several differences exist between agile software development methods and plan-driven

Table 1. Manifesto for agile software development

We are uncovering better ways of developing software by doing it and helping others do it. Through this work we have come to value: ▫ **Individuals and interactions** over processes and tools. ▫ **Working software** over comprehensive documentation. ▫ **Customer collaboration** over contract negotiation. ▫ **Responding to change** over following a plan. That is, while there is value in the items on the right, we value the items on the left more.

Table 2. Several differences between agile and plan-driven software development methods

	Agile Software Development Methods	**Plan-Driven Software Development Methods**
Process orientation	The development process is formulated in terms of activities that all team members apply on a daily basis.	The development process is formulated in terms of stages, in which each team member has one defined role in the process.
Formulation of requirements	Requirements are formulated in a gradual process during which customers and developers improve their understanding of the developed product. This process enables natural evolution.	Requirements are formulated in one of the first stages of the project. Therefore, the cost of implementing a change in requirements increases the later in the process it is introduced.
Customer involvement	Customers are available for discussion, clarifications, etc., in all stages of the software development process.	Primary contact with the customers occurs at the beginning of the development process.

methods. Table 2 summarizes some of these differences.

AQ vs. Plan-Driven QA

In plan-driven software development environments, the main concept related to software quality is quality assurance, which, according to Sommerville (2001), is "The establishment of a framework of organizational procedures and standards which lead to high-quality software" (p. 537). Though this definition inspires an organizational roof for quality assurance processes, in reality, in many software organizations quality assurance is associated with a specific stage of a typical software development process and is usually carried out by the QA people who are not the developers of the code whose quality is being examined.

To illustrate the agile software development approach toward quality, we quote Cockburn (2001), who describes quality as a team characteristic:

Quality may refer to the activities or the work products. In XP, the quality of the team's program is evaluated by examining the source code work product: "All checked-in code must pass unit tests at 100% at all times." The XP team members also evaluate the quality of their activities: Do they hold a stand-up meeting every day? How often do the programmers shift programming partners? How available are the customers for questions? In some cases, quality is given a numerical value, in other cases, a fuzzy value ("I wasn't happy with the team moral on the last iteration") (p. 118).

Table 3. Some differences between AQ and plan-driven QA

	Agile Quality (AQ)	Plan-Driven QA
Who is responsible for software quality?	All development team members	The QA team
When are quality-related topics addressed?	During the entire software development process; quality is one of the primary concerns of the development process	Mainly at the QA/testing stage
Status of quality-related activities relatively to other software development activities	Same as other activities	Low (Cohen, Birkin, Garfield, & Webb, 2004)
Work style	Collaboration between all role holders	Developers and QA people might have conflicts (Cohen et al., 2004)

As can be seen, within the framework of agile software development, quality refers to the *entire team* during the *entire process* of software development and it measures the code as well as the actual activities performed during the development process, both in quantitative and in qualitative terms. Accordingly, the term quality assurance does not appear in agile software development as a specific stage.

In Table 3, we summarize some of the noticeable differences between the attitude toward quality of agile software development methods and of plan-driven methods, as it is manifested in many software organizations.

We note that these previous perspectives are clearly also reflected in the cultures of the two approaches toward software development. While in the context of plan-driven development, conferences are held that are dedicated to QA issues, conferences that target the community of agile software developers subsume all aspects of the development process, including AQ. This difference might, of course, be attributed to the maturity of the plan-driven software development approach; still, the observation is interesting by itself.

TEACHING AGILE SOFTWARE DEVELOPMENT QUALITY

Why Teach QA?

Naturally, software engineers should be educated for quality. The importance of this belief is reflected, for example, in the Software Engineering volume[3] of the Computing Curricula 2001, in which software quality is one of the software engineering education knowledge areas (p. 20), and is described as follows:

Software quality is a pervasive concept that affects, and is affected by all aspects of software development, support, revision, and maintenance. It encompasses the quality of work products developed and/or modified (both intermediate and deliverable work products) and the quality of the work processes used to develop and/or modify the work products. Quality work product attributes include functionality, usability, reliability, safety, security, maintainability, portability, efficiency, performance, and availability. (p. 31)

Table 4. Principles of the software engineering code of ethics and professional practice

1. Public: Software engineers shall act consistently with the public interest.
2. Client and Employer: Software engineers shall act in a manner that is in the best interests of their client and employer, consistent with the public interest.
3. Product: Software engineers shall ensure that their products and related modifications meet the highest professional standards possible.
4. Judgment: Software engineers shall maintain integrity and independence in their professional judgment.
5. Management: Software engineering managers and leaders shall subscribe to and promote an ethical approach to the management of software development and maintenance.
6. Profession: Software engineers shall advance the integrity and reputation of the profession consistent with the public interest.
7. Colleagues: Software engineers shall be fair to and supportive of their colleagues.
8. Self: Software engineers shall participate in lifelong learning regarding the practice of their profession and shall promote an ethical approach to the practice of the profession.

Furthermore, the software engineering code of ethics and professional practice[4], formulated by an IEEE-CS/ACM Joint Task Force, addresses quality issues and outlines how software developers should adhere to ethical behavior. Table 4 presents the eight principles of the Code. Note especially Principle 3, which focuses on quality.

Based on the assumption that the concept of quality should be taught as part of software engineering education, the question that we should ask at this stage is, How should quality be taught? Later in this section, we present our perspective on this matter. We suggest that the nature of the software development methods that inspire a curriculum is usually reflected in the curriculum itself. For example, in traditional software engineering and computer science programs, QA is taught as a separate course, similar to the way in which it is applied in reality in plan-driven software development processes. Based on our teaching experience of agile software development methods, we propose that when teaching the concept of quality is integrated into a software engineering program that is inspired by agile software development, quality-related issues should and are integrated and intertwined in all topics. This idea, as well as others, is illustrated in the next section in which we present the teaching framework we have developed for teaching agile software development and illustrate how AQ integrated naturally into this teaching framework.

Teaching Framework for AQ

This section is the heart of our chapter. In what follows, we introduce our teaching framework, which is composed of nine principles, presented in Table 5 as pedagogical guidelines. Each of the principles is illustrated with respect to the teaching of AQ.

As can be seen, all principles put the learners at the center of the discussion while referring to two main aspects—cognitive and social. Specifically, Principles 1, 2, 3, and 7 emphasize the learning process from a cognitive perspective while Principles 4, 5, 6, 8, and 9 highlight the learning process from a social perspective. We note that this is not a dichotomy, but rather, each principle addresses both aspects to some degree. Accordingly, in what follows, the principles are presented in such an order that enables a gradual

Table 5. Teaching framework

- **Principle 1:** Inspire the agile concept nature.
- **Principle 2:** Let the learners experience the agile concept as much as possible.
- **Principle 3:** Elicit reflection on experience.
- **Principle 4:** Elicit communication.
- **Principle 5:** Encourage diverse viewpoints.
- **Principle 6:** Assign roles to team members.
- **Principle 7:** Be aware of cognitive aspects.
- **Principle 8:** Listen to participants' feelings toward the agile concept.
- **Principle 9:** Emphasize the agile concept in the context of the software world.

mental construction of the learning environment that this teaching framework inspires.

Specifically, for each principle we first describe how it is expressed when agile software development concepts are taught, and then how it is applied in the teaching of AQ.

This presentation style is consistent with our perspective of the teaching of AQ. As mentioned previously, agile software development inspires a development environment in which all activities involved in software development processes are intertwined, and the notion of a production chain is eliminated. Accordingly, when we teach AQ we do not separate it from the teaching of the software development process (in our case, agile software development) but, rather, AQ is taught as part of the software development process in the same spirit in which the entire agile development process is taught.

This section presents, in fact, the application of our teaching framework for software development methods (presented in Dubinsky & Hazzan, 2005 and in Hazzan & Dubinsky, 2006) for the case of AQ. In Dubinsky and Hazzan (2005), we also outline the evolutionary emergence of the teaching framework and describe its implementation in a specific course (including detailed schedule and activities).

Principle 1: Inspire the Agile Concept Nature

This is a meta-principle that integrates several of the principles described later on in this section and, at the same time, is supported by them. It suggests that complex concepts in software development, such as quality or a software development method, should not be lectured about, but rather, their spirit should be inspired. In other words, the teaching of a complex (agile) concept should not be based solely on lecturers but rather, the learning of the main ideas of such concepts is more valuable if a "learning by doing" approach is applied and the (agile) concept is applied, performed, and used by the learners. Such an experience improves the learners experience and skills in the said agile concept, and at the same time, provides the teacher with opportunities to elicit reflection processes.

The application of this principle is expressed by active learning (Silberman, 1996) on which the next principle elaborates, and should take place in an environment that enables the actual performance of the agile concept.

In the case of teaching AQ, this principle implies that the learning occurs in an environment in which it would be natural to illustrate and feel the interrelation between AQ and the other activities that take place in agile software development environments. For example, the extreme programming practice of *whole team*, which states that "a variety of people work together in interlinking ways to make a project more effective" (Beck & Andres, 2004, p. 73), should be applied in order to inspire agile software development. In such software development environments, when the teacher asks the learners to expose and reflect on the relationships between AQ and the other activities, connections between AQ and other activities performed in this environment become clear.

Principle 2: Let the Learners Experience the Agile Concept as Much as Possible

This principle is derived directly from the previous one. In fact, these two principles stem from the importance attributed to the learners' experimental basis, which is essential in learning processes of complex concepts. This assertion stands in line with the constructivist perspective of learning (Davis, Maher, & Noddings, 1990; Confrey, 1995; Kilpatrick, 1987), the origins of which are rooted in Jean Piaget's studies (Piaget, 1977).

Constructivism is a cognitive theory that examines learning processes that lead to mental constructions of knowledge based upon learners' knowledge and experience. According to this approach, learners construct new knowledge by rearranging and refining their existing knowledge (Davis et al., 1990; Smith, diSessa, & Roschelle, 1993). More specifically, the constructivist approach suggests that new knowledge is constructed *gradually,* based on the learner's existing mental structures and in accordance with feedback that the learner receives both from other people with whom he or she interacts and from the different artifacts that constitute the learning environments. In this process, mental structures are developed in steps, each step elaborating on the preceding ones. Naturally, there may also be regressions and blind alleys.

We suggest that quality in general, and AQ in particular, arc complex concepts. Therefore, their gradual learning process should be based on the learners' experience. One way to support and enhance such a gradual mental learning process is to adopt an active-learning teaching approach according to which learners are *active* to the extent that enables a reflective process (which is addressed by another principle later on in this chapter).

We do not claim that lecturing should be absolutely avoided in the process of teaching AQ; in fact, some aspects of AQ can and should be taught by means of lectures. Our experience, however, teaches us that the more learners *experience* AQ and *reflect* upon it, the more they improve their understanding of the essence of the topic, as well as their professional skills.

To illustrate how this principle is applied in the case of AQ, we focus on acceptance tests. Here, active learning is expressed in several ways. First, learners are active in the definition of the software requirements. Second, learners define the acceptance tests and verify that they meet the requirements. Third, they develop the acceptance tests. And fourth, they are guided to reflect both on each individual step and on the entire process. Such a complete process provides learners with a comprehensive message that both highlights each element of the AQ process and at the same time connects each of its elements to the others.

Principle 3: Elicit Reflection on Experience

The importance of introducing reflective processes into software development processes has been already discussed (Hazzan, 2002; Hazzan & Tomayko, 2003). This approach is based on Schön's *Reflective Practitioner* perspective (Schön, 1983, 1987). Indeed, it is well known in the software industry that a reflective person, who learns both from the successes and failures of previous software projects, is more likely to improve his or her own performance in the field (Kerth, 2001).

According to this principle, learners should be encouraged to reflect on their learning processes as well as on different situations in the software development process in which they participated. We note that reflection processes should not be limited to technical issues, but rather should also address feelings, work habits, and social interactions related to the software development processes.

In order to elicit learners' reflective processes, learners should be offered verbal and written means for self-expression. The ability to express one's reflections and impressions gives learners the feeling that their thoughts and feelings are of interest to the instructors. Naturally, such reflective processes might also elicit criticism and complaints. In this spirit, learners should be encouraged to express not only positive ideas, but also negative feelings and suggestions for improvement.

The teaching of AQ is a good opportunity to illustrate this principle since it allows us to address the different facets of AQ. First, we can address the technical aspect of AQ, asking learners to reflect on the actual processes of applying AQ. Specifically, learners can be asked to describe the process they went through, to indicate actions that improved their progress and actions that blocked progress and should be improved, and to suggest how the AQ process itself could be improved. Second, affective aspects can be referred to during the reflection process. For example, learners can be asked to describe their feelings during the AQ process and specifically indicate actions that encouraged them, as well as actions that discouraged them, in their pursuit of the AQ process. Finally, social issues can be addressed in such reflection processes. For example, learners can be asked to indicate what teamwork actions supported the AQ process and which interfered with that process and to suggest how such interactions should be changed so as to support the AQ process. Furthermore, experience learners can be asked to reflect both during the AQ process and after it is completed—processes that Schön calls in-action and on-action reflection, respectively.

Principle 4: Elicit Communication

Communication is a central theme in software development processes. Indeed, the success or failure of software projects is sometimes attributed to communication issues. Accordingly, in all learning situations we aim at fostering learner-learner, as well as learner-teacher communication.

When communication is one of the main ingredients of the learning environment, the idea of knowledge sharing becomes natural. Then, in turn, knowledge sharing reflects back on communication. This principle can be applied very naturally in the context of AQ since it is a multifaceted concept. During the AQ learning process, learners can be asked to identify its different facets (such as, the developer perspective, the customer perspective, its fitness to the organizational culture) and to allocate the learning of its facets to different team members—first learning them, and then subsequently teaching them to the other team members in the stage that follows. In the spirit of agile software development, it is appropriate to assign the different aspects that are to be learned to pairs of learners (rather than to individuals) in order to foster learning processes. When the team members present what they have learned to their teammates, not only do they share their knowledge, but further communication is enhanced.

Another way to foster communication is to use metaphors or "concepts from other worlds." Metaphors are used naturally in our daily life, as well as in educational environments. Generally speaking, metaphors are used in order to understand and experience one specific thing using the terms of another thing (Lakoff & Johnson, 1980; Lawler, 1999). Communication, which is based on the metaphor's concept-world, refers not only to instances in which both concept-worlds correspond to one another, but also to cases in which they do not. If both concept-worlds are identical, the metaphor is not a metaphor of that thing, but rather the thing itself. Specifically, metaphors can be useful even without specifically mentioning the concept of metaphor. For example, the facilitator may say: "Can you suggest another concept-world that may help us understand this unclear issue."

Our experience indicates that learners have no problem suggesting a varied collection of concept-worlds, each highlighting a different aspect of the said problem and together supporting the comprehension of the topic under discussion.

Principle 5: Encourage Diverse Viewpoints

This perspective is based on management theories that assert the added value of diversity (cf. the American Institute for Managing Diversity, http://aimd.org). In the context of agile software development, it is appropriate to start by quoting Beck et al. (2004):

Teams need to bring together a variety of skills, attitudes, and perspectives to see problems and pitfalls, to think of multiple ways to solve problems, and to implement the solutions. Teams need diversity. (p. 29)

We argue that this perspective is correct also with respect to AQ, as explained next.

Naturally, the more diverse a team is, the more diverse the perspectives elicited are. These diverse viewpoints may improve software development processes in general, and the execution of AQ in particular. Specifically, in this context diversity has several benefits. First, learners are exposed to different perspectives that they can use when communicating with people from different sectors and of different opinions. Second, the developed software product itself may be improved because when different perspectives are expressed with respect to a specific topic, the chances that subtle issues will emerge are higher. Consequently, additional factors are considered when decisions are made. Third, the creation process is questioned more when diverse opinions are expressed and, once again, this may result in a more argument-based process based on which specific decisions are made. Finally, we believe that diversity reduces resistance to new ideas and creates an atmosphere of openness toward alternative opinions. In the case of learning AQ, which inspires different work habits than the ones most learners are familiar with, such openness to a different perspective is especially important.

Principle 6: Assign Roles to Team Members

This principle suggests that each team member should have an individual role in addition to the personal development tasks for which he or she is responsible. Based on our agile teaching and research practice, we have identified 12 roles, each of which is related to at least one aspect of software development, several of which are related to AQ (Dubinsky & Hazzan, 2004a). See Table 6.

The role assignment serves as a means for distributing the responsibility for the project progress and quality among all team members. The rationale for this practice stems from the fact that one person (or a small number of practitioners) can not control and handle the great complexity involved in software development projects. When accountability is shared by all team members, each aspect of the entire process is treated by single team member, yet, at the same time, each team member feels personally responsibility for every such aspect. Indeed, both the software project and all team members benefit from this kind of organization.

More specifically, our research shows that the accumulative impact of these roles increases the software quality both from the customer's perspective and from the development perspective, for several reasons. First, the roles address different aspects of the development process (management, customer, code) and together encompass all aspects of a software development process. Second, such a role assignment increases the team members' *commitment* to the project. In order to carry out one's role successfully, each team member must gain a global view of the developed software, in addition to the execution of

Table 6. Roles in agile teams

Role	Description
Leading Group	
Coach	Coordinates and solves group problems, checks the Web forum and responds on a daily basis, leads development sessions.
Tracker	Measures the group progress according to test level and task estimations, manages studio boards, manages the group diary.
Customer Group	
End user	Performs on-going evaluation of the product, collects and processes feedback received from real end-users.
Customer	Tells customer stories, makes decisions pertaining to each iteration, provides feedback, defines acceptance tests.
Acceptance tester	Works with the customer to define and develop acceptance tests, learns and instructs test-driven development.
Maintenance Group	
Presenter	Plans, organizes, and presents iteration presentations, demos, and time schedule allocations.
Documenter	Plans, organizes, and presents project documentation: process documentation, user's guide, and installation instructions.
Installer	Plans and develops an automated installation kit, maintains studio infrastructure.
Code Group	
Designer	Maintains current design, works to simplify design, searches for refactoring tasks and ensures their proper execution.
Unit tester	Learns about unit testing, establishes an automated test suite, guides and supports others in developing unit tests.
Continuous integrator	Establishes an integration environment, publishes rules pertaining to the addition of new code using the test suite.
Code reviewer	Maintains source control, establishes and refines coding standards, guides and manages the team's pair programming.

his or her personal development tasks. This need, in turn, increases one's responsibility toward the development process. Third, the need to perform one's role successfully increases the team members' *involvement* in all parts of the developed software and leads him or her to become familiar with all software parts. If team members have only a limited view and are aware only of their own personal development tasks, they will not be able to perform their personal roles properly. Alternatively, the proposed role scheme supports knowledge sharing, participants' involvement and enhanced performances.

The software quality and the quality of the development process are reflected by three measures that serve as AQ teaching-metrics. The first is the role time measure (RTM). The RTM measures the development-hours/role-hours ratio, or in other words, the time invested in development tasks relative to the time invested in role activities. The second measure is the role communication measure (RCM), which measures the level of communication in the team at each development stage. The third measure is the role management measure (RMM), which measures the level of the project management. Data illustration of these metrics, taken from a specific academic project, can be found in Dubinsky and Hazzan (2004b).

Principle 7: Be Aware of Cognitive Aspects

This principle addresses two issues. The first deals with the idea of inspiring a process of on-going and gradual improvement. The second addresses the fact that software development should be addressed by the individuals involved on different levels of abstraction.

It is clear that software development is a gradual process conducted in stages, each one improving upon those preceding it. In many cases, this improvement takes place in parallel to an improvement in the developers understanding of the developed application. Indeed, this principle is closely derived from the constructivist approach presented in the previous discussion of Principle 2. Accordingly, the learning environment should specifically inspire that feeling of gradual learning and elicit reflection processes when appropriate (cf. Principle 3).

We briefly present two illustrative scenarios that describe how this principle can be applied in practice. When learners try to achieve a consensus with respect to a topic of which their current knowledge is insufficient, the instructor/facilitator should guide them to postpone their final decision until a later stage. Sometimes, the instructor should guide the team to make a temporary decision based on their current knowledge, and explicitly state that in the future they will be able to update, refine, and even change the decision just made. In other cases when learners are deadlocked the moderator/instructor can stop the discussion, reflect on what has transpired, and suggest to move on, explaining that it might make more sense to readdress the issue currently blocking the development progress at a later stage when the learners' background and knowledge can solve the said problem.

As mentioned before, this principle is also related to thinking on different levels of abstraction. In a previous paper (Hazzan & Dubinsky, 2003), we suggested that during the process of software development, developers are required to think on different abstraction levels and to shift between abstraction levels, and explain how several agile practices (such as, refactoring and planning game) support this shift between abstraction level. In other words, developers must shift from a global view of the system (high level of abstraction) to a local, detailed view of the system (low level of abstraction), and vise versa. For example, when trying to understand customers' requirements during the first stage of development, developers should have a global view of the application (high level of abstraction). When coding a specific class, a local perspective (on a lower abstraction level) should be adopted. Obviously, there are many intermediate abstraction levels in between these two levels that programmers should consider during the process of software development. However, knowing how and when to move between different levels of abstraction does not always come naturally, and requires some degree of awareness. For example, a developer may remain at an inappropriate level of abstraction for too long a time, while the problem he or she faces could be solved immediately if the problem were viewed on a different (higher or lower) level of abstraction. The required shift to that different abstraction level might not be made naturally, unless one is aware that this may be a possible step toward a solution.

This principle suggests that instructors or workshop facilitators who teach agile AQ should be aware of the abstraction level on which each stage of each activity is performed. Based on this awareness, they then should decide whether to remain on this abstraction level, or, alternatively, whether there is a need to guide the participants to think in terms of a different level of abstraction. For example, when learners are engaged in design activities and tend to move to details related to the code level, it is important to guide them to stay at the appropriate (higher) level of abstraction. It is further suggested that the instructor or facilitator explicitly highlight the movement between abstraction levels and discuss with the learners the advantages that can be gained from such moves.

We note that the role assignment mentioned in the discussion of Principle 6 can also be viewed as a means to encourage learners to look, think and examine the development process from different abstraction levels. More specifically, if a team member wishes to perform his or her individual role successfully, that is, to lead the project in the direction that the role specifies, he or she must gain a more global (abstract) view of the developed application.

Principle 8: Listen to Participants' Feelings Toward the Agile Concept

The adoption of AQ requires a conceptual change with respect to what a software development process is. In practice, when learners express emotional statements against some aspect of AQ, we propose to take advantage of this opportunity and encourage participants to describe the subject of the said statement as it is manifested in their current software development environment. As it turns out, in many cases these descriptions elicit problems in the currently used approach. Then, we explain how AQ attempts to overcome the problematic issues just raised. For example, when a statement is made against the test-driven development approach, it is a good opportunity to ask the person making this statement to describe the testing process that he or she is currently using. In some cases, this in itself is sufficient: The question highlights the test-driven development approach toward the discussed issue, and consequently, in many cases, the facial expression of the person expressing the objection immediately changes.

In all teaching situations, we propose to try sympathizing with and legitimizing learners' feelings, and being patient until learners start becoming aware of the benefits that can be gained from the new approach. In many cases, learners' objections disappeared in part after a short while. One plausible explanation is that they begin to realize that the new approach might actually support their work and improve the quality of their developed products.

Principle 9: Emphasize the Agile Concept in the Context of the Software World

This principle closes the circle that opened with the first principle—Inspire the nature of the learned concept, in our case—AQ. We believe that part of this inspiration is related to the connections made between the concept taught and the world of software engineering. Since the world of software engineering has witnessed relatively many cases in which new terms emerged and shortly after turned out to be no more than buzzwords, when teaching a new concept that requires developers to adopt a different state of mind, it is preferable to connect the new idea to the world of software development, and in our case, to connect AQ to other agile ideas. This can be done, for example, by presenting the learners with specific problems faced by the software industry (for example, the high rate of software projects that do not fit customer requirements), illustrating how the taught idea may help overcome them. Learners will then, hopefully, feel that, on the one hand, they are being introduced to a new idea that is not detached from the software industry world and is not just a passing fashion, and on the other hand, that the new approach toward quality issues emerged as a timely answer to the needs of the software industry and that it will be useful to them in the future.

In the case of teaching AQ, the need for AQ may be first explained and some problems related to traditional QA processes may be outlined. Such a broad perspective enables learners to understand the place of the agile approach in the software industry in general, and in particular, to observe that AQ is a topic that is still undergoing development.

SUMMARY

The set of principles presented in this chapter aims to establish a teaching framework within which we teach agile software development in general, and AQ in particular. A closer look at the teaching framework reveals that, in fact, its nature is similar to that of agile software development environments. Specifically, as agile software development inspires the notion of a single comprehensive framework in which all activities are performed by all team members in short cycles, with the different activities mutually contributing to one another, the framework described in this chapter also inspires an integrative teaching framework in which all principles should be adhered to at the same time, with different focuses as appropriate. Furthermore, as the assimilation of agile software development takes place in stages, the adoption of this teaching framework should also be carried out gradually, according to the culture of the environments into which the teaching framework is assimilated.

ACKNOWLEDGMENTS

Our thanks are extended to the Technion V.P.R. Fund—B. and the G. Greenberg Research Fund (Ottawa) for their support of this research.

REFERENCES

Beck, K. (2000). *Extreme programming explained: Embrace change.* Boston: Addison-Wesley.

Beck, K., & Andres, C. (2004). *Extreme programming explained: Embrace change* (2nd ed.). Boston: Addison-Wesley.

Boehm, B., & Turner, R. (2004). *Balancing agility and discipline.* Reading, MA: Pearson Education Inc.

Cockburn, A. (2001). *Agile software development.* Boston: Addison-Wesley.

Cohen, C. F., Birkin, S. J., Garfield, M. J., & Webb, H. W. (2004). Managing conflict in software testing, *Communications of the ACM, 47*(1), 76-81.

Confrey J. (1995). A theory of intellectual development. *For the Learning of Mathematics, 15*(2), 36-45.

Davis, R. B., Maher, C. A., & Noddings, N. (1990). Constructivist views on the teaching and learning of mathematics. *Journal for Research in Mathematics Education*, Monograph Number 4, The National Council of Teachers of Mathematics.

Dubinsky, Y., & Hazzan, O. (2004a). Roles in agile software development teams. The *5th International Conference on Extreme Programming and Agile Processes in Software Engineering* (pp. 157-166). Garmisch-Partenkirchen, Germany.

Dubinsky, Y., & Hazzan, O. (2004b). Using a roles scheme to derive software project metrics. *Quantitative Techniques for Software Agile Processes Workshop, Proceedings (and selected for the Post-Proceedings) of SIGSOFT 2004*, Newport Beach, CA.

Dubinsky, Y., & Hazzan, O. (2005). A framework for teaching software development methods. *Computer Science Education, 15*(4), 275-296.

Fowler, M., & Beck, K. (2002). *Planning extreme programming.* Boston.

Hazzan, O. (2002). The reflective practitioner perspective in software engineering education. *The Journal of Systems and Software, 63*(3), 161-171.

Hazzan, O., & Dubinsky, Y. (2003). Bridging cognitive and social chasms in software development using extreme programming. *Proceedings of the 4th International Conference on eXtreme Programming and Agile Processes in Software Engineering* (pp. 47-53). Genova, Italy.

Hazzan, O., & Dubinsky, Y. (2006). Teaching framework for software development methods. Poster presented at the ICSE Educator's Track. *Proceedings of ICSE (International Conference of Software Engineering)* (pp. 703-706), Shanghai, China.

Hazzan, O., & Tomayko, J. (2003). The reflective practitioner perspective in eXtreme programming. *Proceedings of the XP Agile Universe 2003* (pp. 51-61). New Orleans, LA.

Highsmith, J. (2002). *Agile software developments ecosystems.* Boston: Addison-Wesley.

Kerth, N. (2001). *Project retrospective.* New York: Dorset House Publishing.

Kilpatrick, J. (1987). What constructivism might be in mathematics education. In J. C. Bergeron, N. Herscovics, & C. Kieran (Eds.), *Proceedings of the 11th International Conference for the Psychology of Mathematics Education (PME11)* (Vol. I, pp. 3-27). Montréal.

Lakoff, G., & Johnson, M. (1980). *Metaphors we live by.* The University of Chicago Press.

Lawler, J. M. (1999). Metaphors we compute by. In D.J. Hickey (Ed.), *Figures of thought: For college writers.* Mountain View, California: Mayfield Publishing.

Piaget, J. (1977). Problems of equilibration. In M. H. Appel, & L. S. Goldberg (Eds.), *Topics in cognitive development, volume 1: Equilibration: Theory, research, and application* (pp. 3-13). New York: Plenum Press.

Schön, D. A. (1983). *The reflective practitioner.* New York: BasicBooks.

Schön, D. A. (1987). *Educating the reflective practitioner: Toward a new design for teaching and learning in the profession.* San Francisco: Jossey-Bass.

Silberman, M. (1996). *Active learning: 101 strategies to teach any subject.* Boston: Pearson Higher Education.

Sommerville, I. (2001). *Software engineering* (6th ed.). Reading, MA: Addison-Wesley.

Smith, J. P., diSessa, A. A., & Roschelle, J. (1993). Misconceptions reconceived: A constructivist analysis of knowledge in transition. *The Journal of the Learning Sciences, 3*(2), 115-163.

Van Vliet, H. (2000). *Software engineering: Principles and practice.* New York: John Wiley & Sons.

ENDNOTES

1 For further information about our work, please visit our Web site *Agile Software Development Methods and Extreme Programming* (http://edu.technion.ac.il/Courses/cs_methods/eXtremeProgramming/XP_Technion.htm).

2 The term "plan-driven" was introduced by Boehm et al. (2004), who divide the software development methods prevalent today into "agile" and "plan-driven."

3 This volume is part of the Joint Task Force on Computing Curricula 2001 carried out by the Computer Society of the Institute for Electrical and Electronic Engineers (IEEE-CS) and the Association for Computing Machinery (ACM): http://sites.computer.org/ccse/SE2004Volume.pdf

4 ACM Code of Ethics and Professional Conduct: http://www.acm.org/constitution/code.html

Section IV
Agile Methods and Quality: Field Experience

Chapter X
Agile Software Development Quality Assurance:
Agile Project Management, Quality Metrics, and Methodologies

James F. Kile
IBM Corporation, USA

Maheshwar R. Inampudi
IBM Corporation, USA

ABSTRACT

Of great interest to software development professionals is whether the adaptive methods found in agile methodologies can be successfully implemented in a highly disciplined environment and still provide the benefits accorded to fully agile projects. As a general rule, agile software development methodologies have typically been applied to non-critical projects using relatively small project teams where there are vague requirements, a high degree of anticipated change, and no significant availability or performance requirements (Boehm & Turner, 2004). Using agile methods in their pure form for projects requiring either high availability, high performance, or both is considered too risky by many practitioners (Boehm et al., 2004; Paulk, 2001). When one investigates the various agile practices, however, one gets the impression that each may still have value when separated from the whole. This chapter discusses how one team was able to successfully drive software development quality improvements and reduce overall cycle time through the introduction of several individual agile development techniques. Through the use of a common-sense approach to software development, it is shown that the incorporation of individual agile techniques does not have to entail additional risk for projects having higher availability, performance, and quality requirements.

INTRODUCTION

Traditional software development approaches, perhaps best represented by the capability maturity model for software (SW-CMM) (Paulk, Curtis, Chrissis, & Weber, 1993) and its successor the capability maturity model for software integration (CMMI®) (Chrissis, Konrad, & Shrum, 2003), focus on a disciplined approach to software development that is still widely used by organizations as a foundation for project success. While the strength of traditional development methods is their ability to instill process repeatability and standardization, they also require a significant amount of organizational investment to ensure their success. Organizations that have done well using traditional approaches can also fall victim of their success through a strict expectation that history can always be repeated (Zhiying, 2003) when the environment becomes uncertain.

Agile development practices have frequently been presented as revolutionary. There is some evidence, however, that they can offer an alternative common-sense approach when applied to traditional software engineering practices (Paulk, 2001). Perhaps they can be used in part to improve the development processes of projects that do not fit the usual agile model (e.g., critical systems with high availability requirements)? Indeed, it has been suggested that project risk should be the driving factor when choosing between agile and plan-driven methods (Boehm et al., 2004) rather than overall project size or criticality. This implies that certain components of *any* project may be well suited to agility while others may not.

This chapter discusses how agile methods were used on one team to successfully drive software development quality improvements and reduce overall cycle time. This is used as a framework for discussing the impact of agile software development on people, processes, and tools. Though the model project team presented is relatively small (eight people), it has some decidedly non-agile characteristics: It is geographically distributed, it has no co-located developers, the resulting product has high performance and reliability requirements, and the organization's development methodology is decidedly waterfall having gained CMM® Level 5 compliance. Therefore, some of the fundamental paradigms that serve as the basis for successful agile development—extreme programming (Beck & Andres, 2005), for example—do not exist. Nevertheless, they were successfully able to implement several agile practices while maintaining high quality deliverables and reducing cycle time.

Chapter Organization

This chapter is organized as follows:

1. **Background:** Some history is given about our model project team and what led them to investigate agile methods. The concept of using a hybrid plan- and agile-driven method is also introduced.
2. **Approaching Selection:** How did our model project team decide which agile practices to use and which ones to discard? This section discusses the risk-based project management and technical approach used.
3. **Implementation:** This section presents how each selected agile practice was incorporated into the software development process.
4. **Impact:** How did the project team know the implemented agile practices were providing some benefit? This section talks generically about some of the metrics that were used to compare the project to prior projects performed by the same team and the impact the selected methods had on the project.
5. **Future Trends:** A brief discussion about what path will be taken to approach follow-on projects.
6. **Conclusion.**

BACKGROUND

How doth the little busy bee
Improve each shining hour,
And gather honey all the day
From every opening flower!

Isaac Watts, *Divine Songs, 20, Against Idleness and Mischief, 1715*

This chapter introduces several concepts about integrating agile software development techniques into a project that does not have typical agile characteristics. The information provided identifies the conditions that were present at the time our profiled project team began to incorporate agile practices into their decidedly traditional development approach. We begin with a history of a project development team that was unable to meet the expectations of its customer and was unsatisfied with the progress they were making toward meeting their goal of quickly developing a quality product that supported both high availability and high performance. Though the conditions identified are specific to this project and project team, one will most likely find them familiar.

Following an overview of the project and project team, a brief summary is given of some of the existing alternative development methodologies that formed the basis of the team's decision to attempt to integrate agile techniques. Though a short section, it provides some additional insight into the investigatory nature underway to improve the team's results.

This background presents the reader with a contextual overview that will serve to ground the topics discussed later in the chapter. It provides a starting point from which the remaining discussions are based. Because a real project team is being profiled, both the name of the project and the product has been obscured throughout the chapter.

Project and Historical Context

In 2003, a project was undertaken to replace an existing Web application used almost daily by a significant number of individuals (almost 450,000 users). This would not be an ordinary application rewrite, however. When the business analyzed how the product was being used and what its perceived shortcomings were, it became clear that the application needed to be taken in a whole new direction. A project was therefore undertaken to create an entirely new application—one that would incorporate the base functionality of the original application, yet include a significant number of functional improvements, usability enhancements, and external dependencies. This was not the first attempt at replacing this application (a prior attempt ended in failure), but it was certainly the most bold.

This original rewrite project became troubled as requirements seemed to never stabilize and critical milestones were continuously missed. Though it was a medium-sized project with approximately 18 individuals on the development team, there were almost as many analysts, testers, and reviewers and perhaps an equal number of stakeholders. It had the classic characteristics of what Ed Yourdon calls a "death march"—a project in which an unbiased risk assessment would determine that the likelihood of failure is extremely high (Yourdon, 2004). Though the project was considered a success both in delivery and quality, the personal sacrifices were extremely costly. It left the entire team feeling that there needed to be a change in how future projects would be managed and how to adapt to rapid change in the future.

Back to Basics

Interestingly, even though it was recognized that things would have to change, the first change that was made was to be sure the team adhered to what they did not hold fast to the first time: the traditional software engineering life cycle. Though this may seem somewhat counterintuitive, part of the problems faced during the original "death march" project had to do with not maintaining proper control over changes, agreeing to a scope that could not possibly be contained within the time allotted for the project, and not properly evaluating risks and dependencies. In other words, the project team needed to be able to walk before it could run. Since traditional development methodologies were well known and had generally predictable results, they would provide the basis upon which any future process changes would be based.

Several Small Successes

In our experience, it is a common occurrence that several smaller upgrade releases follow large application enhancements or new application implementations—this was no exception. As the project team was re-learning the basics of the software engineering process, there were two opportunities to immediately put it to work and identify which areas were ripe for true improvement. The first was a 2-month cycle of enhancements. It was a small project, but there was still a significant staff on board to complete the work. Unlike the first project, this one adhered to the traditional software engineering process and was successful with respect to schedule, cost, and quality. The business customer was satisfied and somewhat relieved that the delivery was uneventful.

The second project of enhancements was slightly larger in scope, but used less staff and, therefore, had a longer duration. Again, a traditional software development process was followed and the project was successful with regard to schedule, cost, and quality. This second project became a true proof point for the team and was a source of confidence in their abilities. They proved that they could maintain control over these types of projects and deliver high quality work. On the other hand, even though requirements change activity was similar to what occurred in the original project, their ability to control the change was through rejection or re-negotiation—they were unable to accept late changes that might have improved the overall product. A prime example of this was in the area of end user usability. In the traditional software development process being used, ensuring that an application is usable had to be done after the application was essentially complete (during user acceptance). Unfortunately, this meant that there would be no time remaining in the development cycle to address any changes prior to releasing the upgraded product. The implication was that these types of "enhancements" would always have to wait for a follow-on release to be implemented.

The project team also began to recognize that their integration and subsequent testing phases consumed a significant part of the development schedule. Even though the project was generally under control, integration had become a time of heroic sleep deprivation to ensure the schedule was met. It was not the same situation as occurred in the original rewrite project, but it was significant enough that the team recognized that this part of the development process needed to be addressed differently.

Rapidly Changing Business Needs

Though our profiled project team could now be considered successful—after all, they were able to deliver on a set of scope within a defined period of time at a defined cost and with good quality results—the process modifications that they made did not allow them to keep up with the rapidly changing needs of the business. The business could not afford to have 6-9 month development cycles with no changes to the original scope. The releases they sought to put out were time sensitive. They also wanted the amount of functionality contained within each release to remain somewhat flexible. Instead, as new ideas arose, they would be added to a list of ever-increasing "future requirements" or handled as changes that would adjust the end date of the release. There was also the nagging problem of not being able to incorporate usability defect corrections easily into the release where the defects were found without adding a separate "usability" test period with corrections prior to the final user acceptance test period. As it was, they were subjecting users to usability issues that would not be corrected until a follow-on release.

Finally, the business was looking for more out of the team and the team was looking for a better way to do things. Traditional software development practices appeared to be only part of the solution. They had learned to walk, but weren't sure yet how to run.

Delivery Challenges

As more and more functional enhancements were requested by the business, the team began to run into additional delivery challenges. Though quality, cost, and schedule were under control, they were unable to build in the most important features fast enough for the business. In fact, they found that their cycle time to complete a project had actually elongated. In essence, they had traded the chaos of the original schedule for its opposite and found that both didn't really solve their problem (though not being in chaos was infinitely better). They also found that just "following the process" had a chilling effect on their customer relationship. The practice of locking down requirements and stopping change made them appear unresponsive and prone to not delivering value. Though the initial releases following the large rewrite were successful, the sense of pending frustration was becoming palpable. Again, the team recognized that they needed to do something different.

Technical Challenges

Technical challenges do not always get the same attention as other facets of software development when discussing the speed of delivery or quality for the final product, but it was a real concern to our profiled project team. Their customer was not only business-savvy, but had a keen interest in directing which technologies were used. This meant that some portion of the technical solution was imparted to the development team through the requirements gathering process. This could include individual technologies or, in one instance, the final production platform's specifications. To accommodate these types of requirements required a bit of experimentation to ensure they would work. This was something that the traditional development process did not easily support since some of the requirements themselves would derive additional requirements once investigated.

Hybrid Methodologies

Using a hybrid of adaptive and traditional software development methodologies is not as new and radical as it may at first appear. Though some of the concepts related to iterative development and other agile-like techniques can be traced back to at least two decades before the first mass-produced computer was even built (Larman & Basili, 2003), the "traditional" waterfall software development model had gained acceptance by the late 1960s when it was proposed that engineering disciplines should be used to tame wild software schedules (Naur & Randell, 1968). It derives its name from the fact that each step has distinct input and exit criteria that is supported by the surrounding steps (Figure 1). Unfortunately, the model assumes that a project goes through the process only once and that the implementation design is sound (Brooks, 1995).

Soon after being proposed, enhancements started to appear. Over time, several evolutionary techniques were developed as a compliment or replacement to the basic waterfall model including modified waterfalls, evolutionary prototyping, staged delivery, and the spiral model (Boehm, 1988; McConnell, 1996). Each enhancement

Figure 1. A traditional waterfall development model

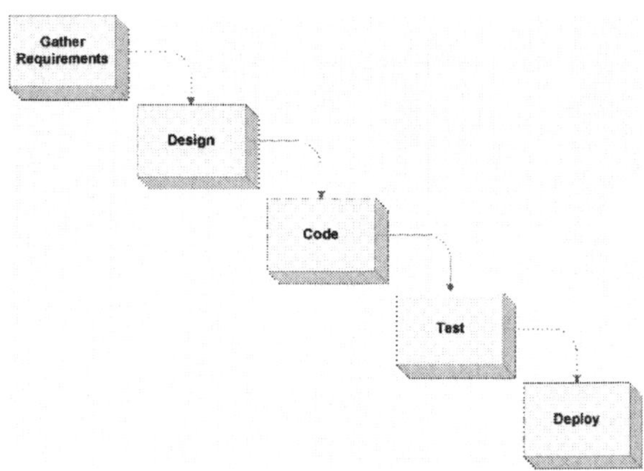

recognized a failing in the original waterfall approach and proceeded to address them within the replacement models.

Why Use a Hybrid?

Why use a hybrid development model and not adopt a single approach? The answer to this question is related to the amount of risk one can afford in their project schedule, cost, and quality. Pure waterfall models operate best with systems that require high reliability and need to be scaleable (McConnell, 1996). Our profiled project team and application has high reliability and high performance as key requirements, but they also have a highly volatile business environment in which the priority of functional enhancements frequently changes.

There is also a bit of a comfort factor in altering something one already understands; One need only learn the new techniques that replaces the original rather than an entirely new process. Over time as new techniques are introduced, the old process will no longer exist in its original form and the organization may be following a totally new methodology—one that meets their needs.

APPROACHING SELECTION

Guess if you can, choose if you dare.

Pierre Corneille, *Héraclius, act IV, sc. IV, 1674*

Great deeds are usually wrought at great risks.

Herodotus, *Histories, VII, 50, c. 485 – c. 425 B. C.*

One of the most difficult things when implementing process change is deciding which changes to make. The entire exercise is a study in risk management since choosing the wrong thing may impact the team's ability to deliver. Recall that after the tumultuous project of 2003, our profiled project team was able to deliver on time, on cost, and at a reasonable level of quality—though there was some room for improvement in the area of quality. Their challenge was to deliver faster and be more adaptable to changes that were brought forward within the development cycle. They recognized that changes needed to be made to make the team's delivery better, but they wanted to be sure that those changes did not undo the predictability they had worked so hard to attain.

The team approached these changes from two perspectives: Project management and technical. From a project management perspective, selected changes would need to be those that would enhance the delivery or quality of the project. From a technical perspective, the changes would need to be reasonable and able to enhance practitioner productivity and delivery. Making changes to one's development process is a unique experience; No two projects are the same. However, there seems to be at least two constants that we will address in the following sections prior to discussing process selection: Fear of change and overcoming that fear.

Fearing Change

Though our profiled project team recognized that there was something that needed to be done to make them a better team that could adapt to changes, deliver more quickly, and produce high quality results, some feared that tinkering with what was working could push them toward the ad hoc development process that they had already rejected. Even though they were not delivering quickly and the customer could not be characterized as completely happy, their projects seemed under control and they were no longer working 90-hour weeks.

The fear of change was manifest in several dimensions for our profiled project team. Each one, though, could be counterbalanced with a fear of not changing. This made for an interesting dance of conflicting emotions around what should be changed and what should be left alone. On one hand, they had proven their competence to their executive management. If they changed the way they do things and failed, they risked something that was tied directly to their self worth. Countering that emotion was the fear of not changing: If their customer was unhappy, the view of their competence may erode regardless.

Overcoming Fear

Fortunately for our profiled project team, their fear of not changing eventually outweighed their fear of change. They were able to recognize that if they did nothing, the situation they would find themselves in would be far worse than if they had not tried at all. Their customer was looking for something new and if the changes could be presented in that light, small bumps in the road may be looked upon more as a learning experience, than failure.

The project management and technical leadership team began to brainstorm. They came up with a plan that would make any change they implemented participative at all levels of the project and conservative so that they could assess the impact and determine if the change was good for the project. Agile practices seemed to make a lot of sense, but a piecemeal approach to change (advocated by those same agile practices) also seemed prudent. They decided that before they implemented any change, they would make sure their customer understood what they were doing and was supportive. In a sense, this approach helped them bridge the chasm between fear of change and the consequences of not changing.

It should be noted that although the project team was able to come to the conclusion that they should change and was able to overcome their fears by making some practical decisions, this was not an easy or quick process. It developed over time and with the help of the relationships they had built with others in the organization.

Process Selection

Implementing changes within any organization takes time and must be participative at all levels to be successful (Manns & Rising, 2005). To overcome the fear of making changes, the team had decided to do it in small steps—a conservative approach that would assist their evaluation of the change when the project was complete. They began by addressing two areas that seemed to cause the most trouble: Requirements prioritization and putting out a version of the release to the customer early so that initial tests—and more importantly usability tests—could be completed in time to provide feedback that could then be incorporated into the code base prior to deployment. Changes would still be controlled, but because there were to be multiple iterations, there would also be multiple integrations and system tests; they would have some flexibility to incorporate small changes from the first cycle into the second assuming they could keep the quality of the release high and they planned enough time for these anticipated changes.

When the team found they had some success (see "Impact") with their initial changes, they became more emboldened. They suggested and implemented more changes. We discuss the areas that were addressed earliest in the next several sections. They are presented along with the reasoning behind each so that the reader can understand why each was chosen by the project team. Later in the chapter, a discussion ensues about how each practice was specifically implemented from a technical standpoint and the cycle time and quality impacts of each.

Prioritizing Requirements

One of the most difficult things facing our profiled project team was their joint ability with their customer to prioritize their requirements. On any given day, the priority may change. What seemed to be missing was a way to quantify the requirements in a manner that would permit a reasonable prioritization. In some cases, a requirement may be highly desired, but its cost would make implementation prohibitive. In other cases, a requirement may be somewhat desired, but its cost would make implementation highly desirable. A process was needed to quickly assess requirements and have the customer prioritize them so that the team was always aware of what features and functions were desired next.

Iterative Development

Partially to address their overall product quality and to gain early feedback on how a release was progressing, the team decided that some form of iterative development should be implemented. Creating products iteratively goes back to an invention theory from the 1920s and 1930s (Larman et al., 2003). It is a proven technique for addressing product quality and change. As you will see, the team's first foray into iterative development was only partially successful and required some additional process changes.

Continuous Integration

Perhaps the most frustrating part of the development process for our profiled project team was the "integration" cycle. This was where the system was put together so that it could be functionally tested as a whole. Part of the frustration with this process was that there was no easy way to see the entire system in operation from end to end without going through a lot of tedious build steps. To address this, the team decided that they would need to automate their builds and would need to permit any team member to create a locally running version of the full system at any time.

Addressing integration took on additional importance with respect to iterative development. If the team wished to create rapid iterations in the future, they could not do so without addressing the integration and build process.

Automation

One area the team thought they could gain improvements in both quality and cycle time was in the area of automation. It has long been understood that design and code inspections could significantly and positively impact the quality of a product, but the time to perform the inspections could be prohibitive for a large product. Indeed, testing also fell into this same category—large benefit, but time consuming. To address the latter concerns, the team identified automating critical reviews and testing as one of their top priorities. Tools such as JUnit, JTest, Rational Performance Tester, Findbugs (http://findbugs.sourceforge.net/), SA4J (http://www.alphaworks.ibm.com/tech/sa4j), and Parasoft's WebKing would be used (and re-used) to reduce their cycle time while improving quality.

IMPLEMENTATION

For the things we have to learn before we can do them, we learn by doing them.

Aristotle, *Nicomachean Ethics, II, 1, ca. 384-322 B. C.*

Deciding which processes to alter as discussed in "Approaching Selection" was only one facet of implementing change. Each area that was selected required a corresponding implementation action (or actions). This section of our chapter focuses on those actions that were taken to address overall quality and cycle time. Of interest are some of the reasons why certain actions were taken. As you will see, the way agility was worked into our profiled project team's project can serve as a model for other project using a hybrid development methodology where teams are looking for incremental or evolutionary (rather than revolutionary) process improvements.

Improving Quality

Perhaps one of the most vexing problems faced after the tumultuous 2003 project and even in the small step enhancement projects undertaken in 2004, was the fact that defects were being discovered and corrected late in the development cycle when they were most time consuming and most expensive to correct. Adjusting the defect detection curve such that it afforded an earlier indication into what was wrong and provided the ability to make early corrections was considered of paramount importance to improving overall code and product quality.

After taking a retrospective look back at how the product itself was developed, it became clear that not everything had been created in a manner that would be considered optimal. There were architectural and design flaws that were not necessarily apparent when the original version of the application was created, but began to impose limitations on development as enhancements were being considered—limitations that had the result of increasing the amount of time and money it would take to make those enhancements.

In addition, the original project team that worked on the first version of the product in 2003 was quite large. Due to the ad hoc nature of that project, no coding standards had been defined or adhered to. This meant that each functional area of the application was implemented and behaved differently. In effect, the application had internal vertical silos of discrete functionality.

Changes surrounding the quality of the application needed to address each of these issues: Defect detection and correction, architectural and design dependencies, and the silo effect of independently created functions. The sections that follow provide a general overview of what was implemented to address each of these concerns. We begin with a discussion about the project's quality management plan. From there, we introduce the concept of "technical stories" as a way the project team codified the refactoring of existing inefficient architectural and design constructs. Next is a description of what was done to move the defect detection and correction curve earlier in the development cycle. This is followed by a description of some of the methods and tools that would be used to enforce best coding practices. Finally, an explanation of how continuous integration might be used to improve overall code quality is given.

Quality Management Plan

Being a traditional waterfall project with a structured approach to development meant that our profiled project team had implemented a quality management plan for each of their projects. Typically, this plan would identify, along industry standard lines, the percentage and aggregated number of defects that one could expect by project phase, how those defects would be detected (e.g., inspection, testing, etc.) and how they would be removed.

Rather than discard the quality management plan, the team thought it important to take the time to update it with the strategy they would employ to address the quality of the product. Though such a document may be derided as defying the "barely sufficient" nature of an agile project, the team found it useful to document an internal goal for the overall detection and correction of defects and the strategy they were going to use for their early elimination from the product. This document also gave them a baseline from which they could measure their intended results with their actual results.

The quality management plan, therefore, became a document that identified the goals the team would strive to achieve and the techniques they would employ to integrate agile practices into their development process. It no longer specified only industry standard information to which the project would attempt to comply, but a much higher standard to which the project team wished to be held. These changes are evident in each of he implementation practices outlined in the following sections. Each, however, was first identified as part of the project's quality management plan.

Technical Stories

One thing that was identified quickly by the profiled project team was that innovation could often be introduced into a project that would also satisfy a business requirement. In other words, the way new function was added began to have both a technical component in addition to the business component. These so-called "technical stories" became part of the requirements gathered after each release and, later on, iteration. They were influenced by a retrospective look at what went well and what did not go so well during each development cycle. As a result of these reflections, the architecture of the application was reduced and simplified through refactoring. This had the net effect of reducing the cost of ownership and change while improving the overall quality of the application by consolidating change points. The approach the team took is similar to the "user stories" concept in extreme programming.

A few possible outcomes of these "technology stories" include:

- Cost reduction as a result of simplification.
- Architecture simplification through refactoring.
- Improvement in the throughput or performance of an individual application module or area.
- Architectural changes that will re-align the application with the long-term strategy of the organization.

Defect Detection and Correction

The continuous feedback provided to the development team through the use of iterative development and continuous integration paired with automation and reuse supplied them with the opportunity to detect and correct defects earlier in the development cycle and improve overall quality. Recalling some of the difficulties the project team had with late usability changes and the difficulty they had integrating the system, two practices were introduced: Test case reuse and test case automation.

Test Case Reuse

When a project is undertaken to enhance an existing product, a common scenario that many developers face is the re-introduction of defects from prior releases when a new feature is added without understanding the overall purpose of the module (or function). One way to combat this is to retain the unit and functional test cases from release to release and execute them prior to and

Figure 2. Technology-based proposals in release planning

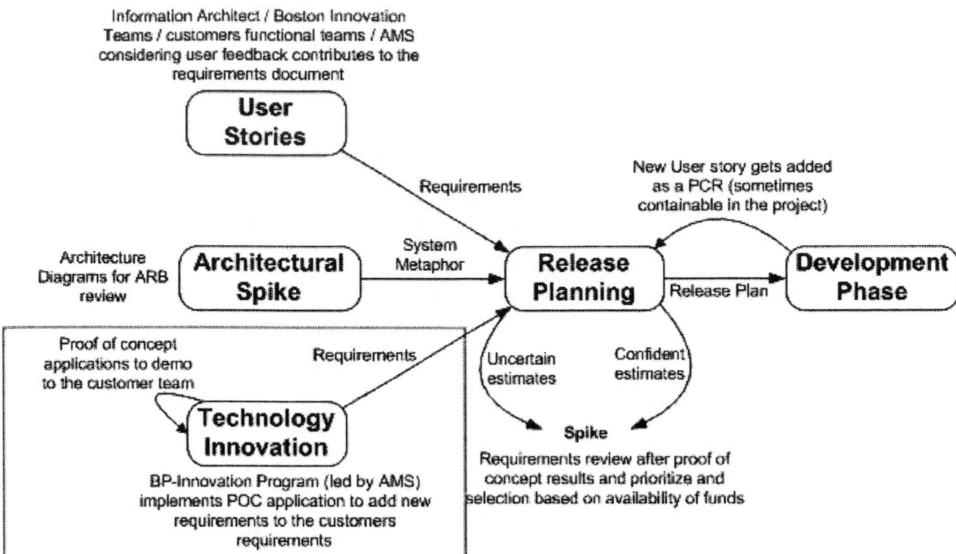

Figure 3. Test case reuse process

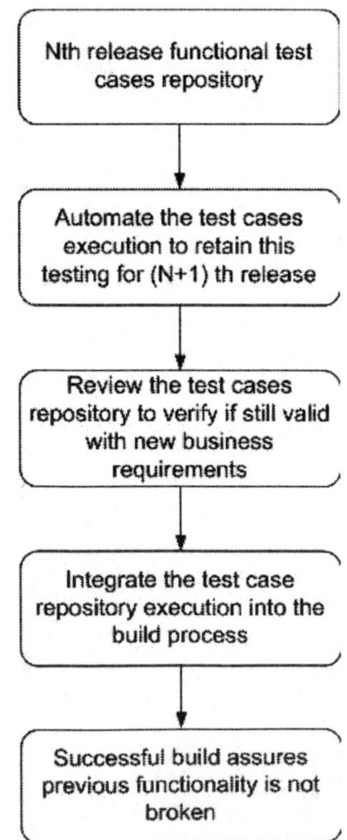

during each build of the system. By integrating execution of the test cases with the build process, one can be assured that existing functionality is not compromised by changes made to the product; Either the new changes are incorrect or the test case is no longer valid. This reduces the number of undetected defects in a release and improves the overall quality of the application. Instead of finding issues during integration or system testing, they can be found and corrected prior to or during each system build. The theory is that the closer the defect is detected to when the change is made, the easier it will be to recall what was changed and fix it. An example of the process followed appears in Figure 3.

Automated Test Case Execution

Agile principles encourage developers to adopt test-driven development. Whether a project follows a pure agile approach, a hybrid approach (as was used here), or a traditional methodology, there

Figure 4. Automated functional test cases using Rational Functional Tester

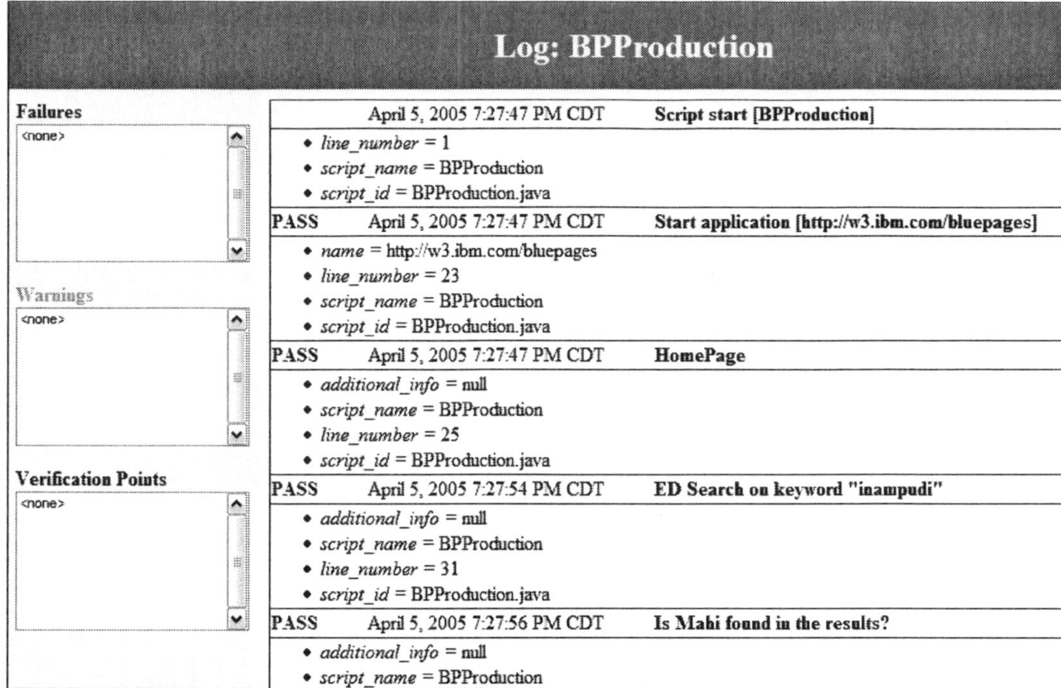

is value in testing code in an automated fashion at the unit and function level. Retaining these test cases so that all developers can execute them in an automated fashion to ensure that their changes do not break the system is an agile principle that was implemented for this project team to measure project progress, address interim code quality, and assist in the development of new classes or methods. It should be noted that since these test cases were being built upon an existing product that did not have them initially, they were first built against those areas that required change. Those cases remained available as subsequent projects were undertaken.

Two tools were used to automate test case execution. The first, not surprisingly, was JUnit for unit testing. For functional testing, IBM's Rational Function Tester was used. This latter tool easily integrates with the build process and provides an automated functional regression testing platform for client-based and Web-based applications. A sample report appears in Figure 4.

Enforce Coding Practices

One area of quality that is oftentimes not addressed by a project team is the way code will be written. Documenting the coding standards up front is helpful, but it will not ensure that an individual will not violate the project's standards or best coding practices in general. Rather than implement a series of manual code inspections, several tools were implemented to ensure best practice compliance.

Automated Code Reviewers

Tools such as Parasoft's JTest, RAD Code Reviewer, and WebKing can be plugged right into the developer's IDE. They help ensure that code is written according to a standard the team has

Figure 5. Example automated code review rules

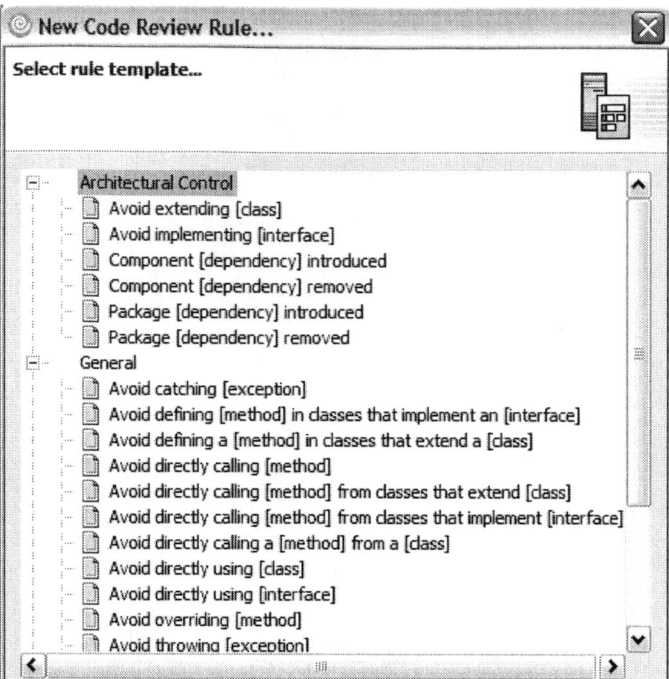

set. They also can catch common mistakes and identify problem areas that may need to be addressed. Each developer on the project team was required to install the plug-ins into their development environment and execute the review process prior to checking the code in or integrating it into the system build. An example of some of the rules used appears in Figure 5.

Tools such as IBM's Rational Application Developer Code review tool can be used to show the details and the nature of a violation including the class name and the line number of the code where the violation occurred (see Figure 6).

Automated Static and Stability Analysis

Static analysis tools such as Findbugs (http://findbugs.sourceforge.net/) and Structural Analyzer for Java (SA4J) (http://www.alphaworks.ibm.com/tech/sa4j) can be embedded into the build process to verify the quality of the build. These tools produce reports for the development team that help them understand potential run time defects either due to the way something was implemented or by finding architectural anti-patterns that can reduce the application's stability in the production environment.

"Continuous" Integration

One of the extreme programming practices that our profiled project team readily adopted was the concept of continuous integration. Recall that one of the most difficult project activities was the integration of created components into a functioning whole. It was felt that if the integration of the application could become more continuous—more agile—it would be possible to see the system working earlier, identify and remove integration defects in closer proximity to when they were introduced, and enforce the view that the system must always be kept in an operational state.

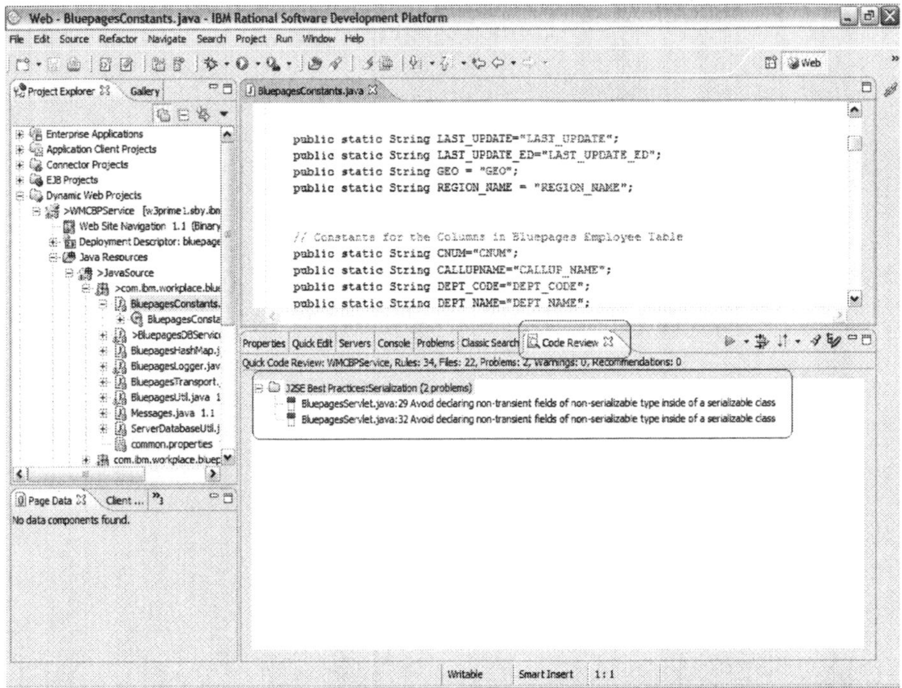

Figure 6. Example code review automation (IBM Rational Code Reviewer)

Automating the Build

The primary barrier to continuous integration was an onerous build process. One of the technical stories, therefore, was to automate the product build such that it could be started by any developer in their own workspace and in the integration environment (under the proper controls). Apache's ANT (http://ant.apache.org), as the de facto standard for build automation, would be used as the foundation for this automated build process. In addition to automating the build itself, the script would also incorporate several of the code verification steps identified earlier: functional analysis, structural analysis, functional test verification, coding practices, etc.

The Build Process

The following process provides a general overview of the steps to be followed by the automated build process identified in several technical stories for the application.

- Pull the latest checked-in source software from the library control system (CVS).
- Execute automated code analysis tools on the extracted code to verify the code's look and feel and identify any anti-patters violations of best practices in the code base.
- Build an EAR package and execute all existing JUnit test cases against the code package and push the results to a build status Web page.
- Install the application into the runtime environment.
- Execute the automated functional test cases using Rational Functional Tester and publish the results to the build status Web page.
- Execute an overall architectural quality check using structural analysis tools (e.g., SA4J).

Figure 7. Example static analysis report

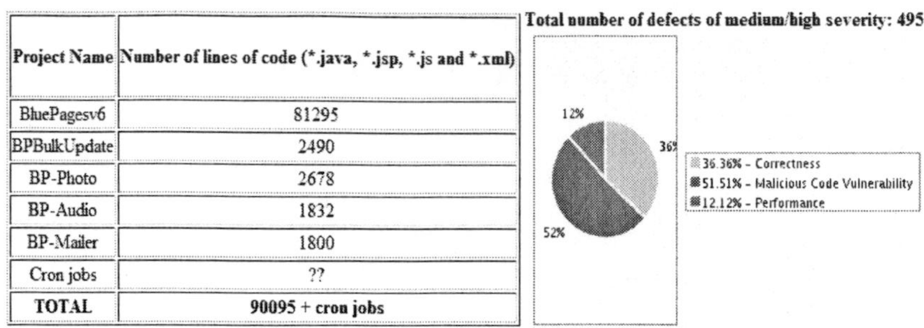

Code Analysis (Findbugs) Summary Report

Summary Analysis Generated at: *Wednesday, April 27, 2005 9:19:04 AM CDT*

Project Name	Number of lines of code (*.java, *.jsp, *.js and *.xml)
BluePagesv6	81295
BPBulkUpdate	2490
BP-Photo	2678
BP-Audio	1832
BP-Mailer	1800
Cron jobs	??
TOTAL	**90095 + cron jobs**

Type Checked	Count	Bugs	Percentage
Outer Classes	227	1261	555.51%
Inner Classes	23	29	126.09%
Interfaces	5	0	0.00%
Total	255	1275	500.00%

A graphical representation of this process appears in Figure 8.

Mapping the Build Framework to the Development Process

Every time a developer adds code to the system either as part of an iteration or as part of a release, the overall integration with existing code (and code written by others) is taken care of by the build process. This process is depicted in Figure 9.

IMPACT

Nothing quite new is perfect.

Marcus Tullius Cicero, *Brutus, 71, c. 106 B.C.-43 B.C.*

What was the overall impact of the changes that were made? By and large, the impact was positive. The team proved they could successfully integrate agile techniques into a traditional development process. What follows is a summary of some of the results. It should be noted that these are results from one team and that the experiment would need to be expanded to others to assess its validity in a broader context. Regardless, some of the results are rather remarkable.

Requirements Prioritization: Time Boxing the Schedule

As identified in the "Approaching Selection" section, instead of beginning with a fixed set of requirements from which a project sizing and project plan was derived, the most important requirements were identified and given a rough sizing. Based upon this rough sizing, the requirements were re-ordered. The schedule was broken down into discrete "time boxes" that dictated how much would be spent and how many features could be contained within a particular iteration or a project. Anything that could not fit would be re-prioritized into a follow-on project (or iteration). This method permitted the customer to introduce what they considered the most important features into the product and regularly deliver function to

Figure 8. Automated build tools stack

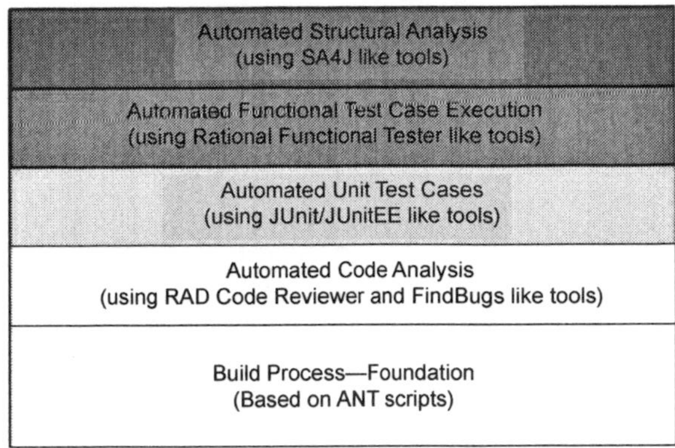

Figure 9. Continuous integration using automated tools

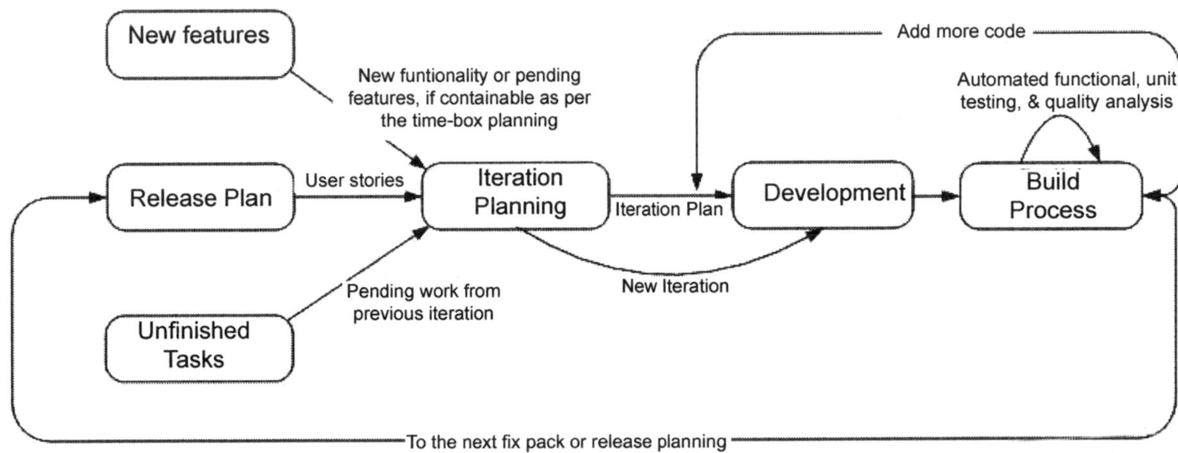

the business. Since the estimation at the beginning was by necessity a rough order of magnitude, as the team began the work additional adjustments would be made to the scope. If the size of a feature grew beyond what could be contained within a project, it would be discarded (or, if truly important, the date would change). If the size of a feature was smaller than what was originally anticipated, additional features would be added from the prioritized list (see Figure 10).

Using this approach, customer satisfaction increased from essentially failing on the first project to 97.3% on the first project that used this approach—they recognized that the project team was willing to help them meet their goals and change the way they do things in support of those goals. (Note: This satisfaction rating was for the entire release and most likely reflects the improvement in quality—discussed later—as well as the way requirements were prioritized.)

Iterative Development

Breaking each release into smaller iterations had a three-fold impact on the projects. First, the customer was able to see the results of built-in features earlier. This allowed them to re-assess the priority of the remaining requirements against

Figure 10. Time boxed requirements prioritization

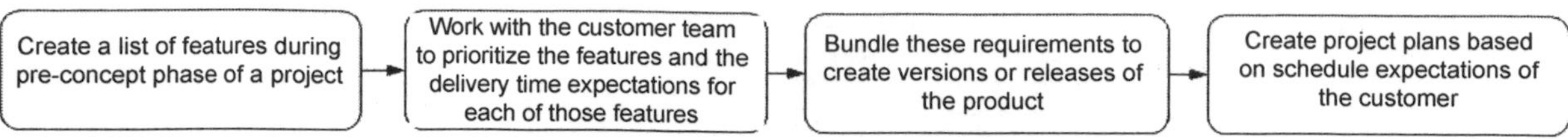

changes that they may wish to implement. From a customer satisfaction perspective, the team was beginning to be seen as much more flexible—yet, the project was still under control.

The second area of positive impact area was the team's ability to have usability testing done on the parts of the application as they were produced. This meant that any usability defects identified could be rolled in to the next iteration and would be in the final build rather than waiting six months for a follow-on project to be commissioned. This had the net effect of improving end user satisfaction to above 85%—which was a significant improvement from 76% given the size of the population.

The third area of impact was in the quality of the final product. Since the system was being "put together" more frequently (also see the discussion on continuous integration results), the amount of time spent cleaning up integration errors was significantly reduced. While the initial 2003 project had greater than 400 user acceptance defects, less than a year later the user acceptance phase for all of the iterations combined had three defects (one of severity 3 and two of severity 4).

As we mentioned, not everything was positive. The way the team initially implemented iterations was not as clean as it could be. They were rushing at the end of the first iteration to get function working. This lesson learned was built into their second project—they more discretely planned the iterations so that they would have enough time to put the system together prior to the interim review by the customer. Interestingly, when they automated the build and began to use a more continuous form of integration, this additional planning was no longer required.

Continuous Integration

Perhaps one of the biggest gains the project team saw from a quality perspective was as a result of implementing their version of continuous integration. As previously discussed, this involved automating the build process such that testing occurred at each run and the system always remained in a workable state. Creating the build process cost approximately 100 labor hours to the team. The amount of time saved in integration and integration testing, however, was 300 hours and almost two weeks cycle time. On the project it was implemented in, the additional time was used to contain additional features that originally did not fit in the planned time boxes. For future projects, the additional time will be factored into the schedule as available for general development.

Automation

Although automation took on various forms including the creation of the automated build used for continuous integration, there were some additional positive impacts to cost and quality. For example, even though there was a cost to modifying automated test cases from release to release, that cost was minimal compared to creating a new test suite each time or executing all of the test cases manually. Some interesting statistics were gathered from project to project. The original project in 2003 used 12.7% of its overall budget (hours) to conduct functional and system testing (not user acceptance testing where most of the defects were eventually found). Through automation and reuse of the test case library, the two subsequent similarly sized projects consumed 5.8% and 5.2% of their budget on function and system testing respectively.

Recall that automation also added several tools that checked the stability and quality of the code. Perhaps the best measure of impact on the project is that after incorporating the recommendations for coding standard best practices and addressing structural weaknesses, the amount of time required to maintain it was reduced by almost 10%. In a world where operation budgets are constrained, this was considered a significant under run. Most of it related to the reduced amount of call support from people who were having trouble using the application or finding obscure errors that had not been caught in the project team's own testing.

FUTURE TRENDS

In our opinion, the trend toward using agile software development practices in general and as a way to enhance the quality of products developed using traditional practices will continue. As with the profiled project team used as the basis for this chapter, we also see a trend toward using risk to identify which practices may work best in a particular environment. That will mean that projects that are not thought of as being able to easily use agile practices to enhance quality or reduce cycle time and cost today—such as those with high availability requirements or high performance requirements—may have an agile component in the future.

Smaller and More Frequent Releases

Developing a product in a piecemeal fashion predates computing by several decades. The concept of creating incremental releases to products initially grew from a desire to improve quality (Larman et al., 2003). Recent evidence has continues to show that smaller, more frequent releases have a positive impact on the overall quality of a software development project (see Madsen, 2005, for example). Several "heavier" methodologies such as the rational unified process always embraced iterations and even that has had its share of agile practices introduced as process improvements (Ambler, 2006). We expect this trend toward smaller, incremental releases with agile components to continue.

Reviews

Another future trend in agile quality management seems to be the return of peer reviews. Agile practices typically rely on up front test cases to ensure quality, but some of the current literature indicates that peer reviews still play an important role in software development. Some recent research has been conducted on focusing reviews on the most important aspects of a particular project based upon risk and the perceived value of a particular review (Lee & Boehm, 2005). This suggests that reviews themselves may also be moving toward a sufficiency model similar to agile. It will be interesting to see if a review structure will appear as part of pure agile practices.

More Hybrids

As with our profiled project team, not everyone is willing or able to move to completely agile approaches for their software development either due to perceived complexity or performance and availability requirements. We believe that the evolutionary introduction of agile practices into traditional organizations will continue, but alterations may be required for an organization to derive value as in Svensson and Host (2005). Perhaps the largest focus area in the next couple of years will be in project management. Project managers will need to not only drive the implementation of agile practices, but also need to understand their impact on their project(s) (Coram & Bohner, 2005). In all of these cases, we believe risk will most likely be the deciding factor for when agile methods are used and when they are not.

CONCLUSION

This chapter discussed how one team was able to successfully drive software development quality improvements while reducing overall cycle time through the introduction of several individual agile development techniques. Through piecemeal change to their existing development processes, they were able to make significant improvements over time. This common-sense approach to software development showed that the incorporation of agile techniques does not have to entail additional risks for projects that have high availability, performance, and quality requirements.

REFERENCES

Ambler, S. W. (2006). *The agile edge: Unified and agile.* Software Development Retrieved January 8, 2006, from http://www.sdmagazine.com/documents/s=9947/sdm0601g/0601g.html

Beck, K., & Andres, C. (2005). *Extreme programming explained: Embrace change* (2nd ed.). Boston: Addison-Wesley.

Boehm, B., & Turner, R. (2004). *Balancing agility and discipline: A guide for the perplexed.* Boston: Addison-Wesley.

Boehm, B. W. (1988). A spiral model of software development and enhancement. *Computer, 21*(5), 61-72.

Brooks, F. P. (1995). *The mythical man-month* (Anniversary Edition). Boston: Addison-Wesley.

Chrissis, M. B., Konrad, M., & Shrum, S. (2003). *CMMI: Guidelines for process integration and product improvement.* Boston: Addison-Wesley.

Coram, M., & Bohner, S. (2005, April 3-8). *The impact of agile methods on software project management.* Paper presented at the 12th IEEE International Conference and Workshops on the Engineering of Computer-Based Systems (ECBS'05), Greenbelt, Maryland, USA.

Larman, C., & Basili, V. R. (2003). Iterative and incremental development: A brief history. *Computer, 36*(6), 47-56.

Lee, K., & Boehm, B. (2005, May 17). *Value-based quality processes and results.* Paper presented at the 3rd Workshop on Software Quality (3-WoSQ), St. Louis, Missouri.

Madsen, K. (2005, October 16-20). *Agility vs. stability at a successful start-up: Steps to progress amidst chaos and change.* Paper presented at the 20th Annual ACM SIGPLAN Conference on Object-Oriented Programming, Systems, Languages, and Applications (OOPSLA '05), San Diego, CA,.

Manns, M. L., & Rising, L. (2005). *Fearless change: Patterns for introducing new ideas.* Boston: Addison-Wesley.

McConnell, S. (1996). *Rapid development: Taming wild software schedules.* Redmond, Washington: Microsoft Press.

Naur, P., & Randell, B. (1968, October 7-11). *Software engineering: Report on a Conference Sponsored by the NATO Science Committee.* Paper presented at the 1st NATO Software Engineering Conference, Garmisch, Germany.

Paulk, M. C. (2001). Extreme programming from a CMM perspective. *IEEE Software, 18*(6), 19-26.

Paulk, M. C., Curtis, B., Chrissis, M. B., & Weber, C. V. (1993). *Capability maturity model for software, Version 1.1. Software Engineering Institute: Capability Maturity Modeling,* 82.

Svensson, H., & Host, M. (2005, March 21-23). *Introducing an agile process in a software maintenance and evolution organization.* Paper presented at the 9th European Conference on Software Maintenance and Reengineering (CSMR'05), Manchester, UK.

Yourdon, E. (2004). *Death march* (2nd ed.). Upper Saddle River, NJ: Prentice Hall.

Zhiying, Z. (2003). CMM in uncertain environments. *Communications of the ACM, 46*(8), 115-119.

Chapter XI
Test-Driven Development: An Agile Practice to Ensure Quality is Built from the Beginning

Scott Mark
Medtronic, Inc.

ABSTRACT

This chapter describes the practice of test-driven development (TDD) and its impact on the overall culture of quality in an organization based on the author's experience introducing TDD into four existing development projects in an industrial setting. The basic concepts of TDD are explored from an industry practitioner's perspective before elaborating on the benefits and challenges of adopting TDD within a development organization. The author observed that TDD was well-received by team members, and believes that other teams will have this same experience if they are prepared to evaluate their own experiences and address the challenges.

INTRODUCTION

The spirit of agile methods begins with the promise that whatever software components are built will be of high quality. Agile methods move quality assurance upstream in the software development process, and the most relevant of these methods is the principle of test-driven development (TDD). The essence of TDD is that quality assurance methods are not a sieve through which application code is pushed at the end of a long, drawn out development process. Rather, the development cycle begins with capturing test cases as executable system components themselves. These testing components are then used to drive the development process and deliver components that, by definition, satisfy the quality requirements as they are set out in the test cases. In its purest form, the developer[1] begins by writing a test case that fails and then proceeds to implement the functionality that causes the test to succeed. When this practice is followed carefully, the code that becomes the final product is guaranteed to pass all currently identified test cases.

The goal of this chapter is to explain the changes to the traditional development process in order to drive it with quality assurance, and illustrate the overall impacts on software quality, process velocity, and developer productivity. The perspectives on TDD presented in this chapter are based on the author's experience introducing these techniques on four Web application development projects in a large enterprise setting. These projects will be described and the teams' good and bad experiences with TDD will be explored. The intention of this chapter is to share the experiences, both good and bad, of these teams while using TDD so that other practitioners can anticipate or evaluate similar effects in their own environments.

WHAT IS TEST-DRIVEN DEVELOPMENT (TDD)?

Test-driven development is the practice of implementing executable tests before implementing functional components, and using the activity of testing to propel the implementation of functional components. For purposes of this discussion, the essential components of the test-driven development practice are the following:

- Tests are authored by the developer before implementation.
- Tests are "easily" executed by the developer working on the implementation.
- Tests are at the unit- or component-level.

Tests are Authored by the Developer before Implementation

TDD is a quality improvement process that ultimately is a form of organizational change. A key aspect of this is the transformation of every developer into a tester. Organizations that have separate roles for authoring tests have not completed this transformation—testing will remain a back-and-forth process of transfer. So a requirement of TDD is that the developer who will be implementing functionality begins by writing the test to verify the implementation.

Tests are "Easily" Executed by the Developer Working on the Implementation

This is of course a very subjective metric, but is a key requirement nonetheless. The core of the TDD practice is that running tests is part of the moment-to-moment development process. For TDD purposes, tests should be kept at a practical level of granularity with a consideration toward execution time. Low execution times ensure that these tests can in practice be run frequently and continuously during the work of implementation.

Tests are at the Unit- or Component-Level

There are various levels of testing within the larger landscape of software quality assurance such as unit, functional, system, integration, and user acceptance testing. It is certainly not the goal of TDD to address all of these aspects of testing. TDD promises to increase the amount of quality that is built-in from the start, and encourages developers to think upfront about testability. This is achieved by testing the aspects immediately in front of the developer at the unit- and component-levels of implementation. A unit test is a test that focuses on a given implementation construct (a .java file in Java, a .c file in C/C++, etc.). A component test is a test that focuses on an atomic system component, such as an interface in Java, that might front a number of implementation constructs in the implementation.

As we proceed to higher levels of granularity throughout the system, test-driven development starts to dissolve into more integration-oriented testing methods. As more and more layers are integrated into the working, testable system, the setup and cycle times of execution increases to the point where some of the benefits of test-driven development diminish. Some of these tests can be automated and some can be written before the implementation. Automated tests written at the integration level often drive refactoring more than they drive initial development. So integration-level tests are not the primary focus of TDD practices. A key motivation of TDD is to ensure that unit-level defects do not seep into the integration phase where they consume valuable time and resources.

THE COMPONENTS OF TEST-DRIVEN DEVELOPMENT

This section will more thoroughly define the technical components or building blocks of TDD. There are four major components (see Figure 1) to a TDD framework:

- Test cases.
- Test suites.
- Test fixtures.
- Test harnesses.

Test Cases

The test case is the primary building block for TDD. The test case is the initial code that the developer writes to exercise an interface and drive implementation. Test cases contain the basic assertions that indicate expected behavior from the implementation. The most important code written for TDD will be written in the test cases, and this code will drive the interface of the implementation.

Test Fixtures

Test fixtures provide support services for tests in the way of setting up proper pre- and post-conditions. By far the most common use of test fixtures is to prepare and dispose of necessary test data against which tests will be run. Fixtures can be responsible for creating test objects, establishing any required dependencies, or mocking.

Test Suites

A test suite is a collection of test cases that are intended to be executed together as a unit. Test cases might be organized into suites according to functional groups, application tiers, or common fixture needs. There is often a hierarchical arrangement of test suites with suites containing other suites and so on. Suites are used mainly for convenience and to indicate which tests are intended to be run as a group.

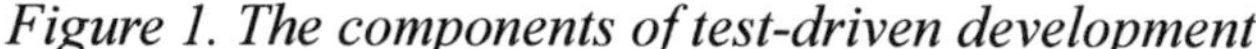
Figure 1. The components of test-driven development

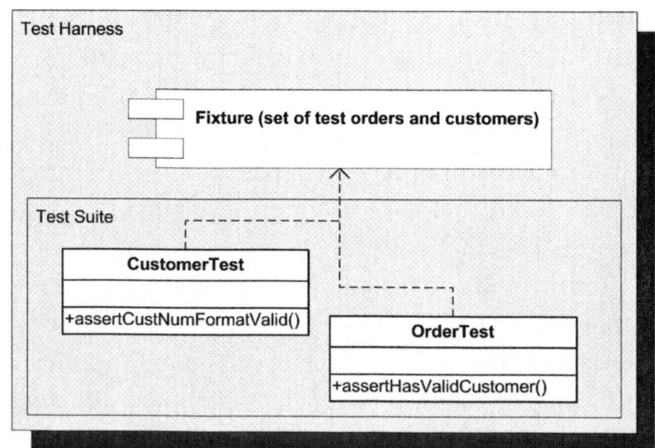

Test Harnesses

The test harness is the highest level component in the TDD framework. The harness is the foundation that ties all of the other testing components into an executable set. At a minimum, a harness provides an execution environment in which tests can be run and output captured. Harnesses might also include features to ease the process of TDD, such as supporting services for more convenient fixture building or formatted reporting of test execution results. By far the most popular harnesses for TDD are the family of xUnit frameworks, such as JUnit, NUnit, and CPPUnit, which were based on the original SUnit framework created by Kent Beck for Smalltalk (Beck, 1999).

WHAT ARE ALTERNATIVES TO TEST-DRIVEN DEVELOPMENT?

Software testing has typically been treated as a topic to be addressed late in the development cycle by many project methodologies. Often after software components are considered "complete" by software development teams, they are delivered to a testing team, at which time an intensive testing phase begins. In traditional "waterfall" methodologies, the testing phase is literally a protracted phase that is ultimately considered a pre-deployment phase. This phase is essentially the combination of all forms of testing (unit, functional, integration) in a single phase that occurs after the conclusion of most major development across the scope of an entire system. In more iterative methodologies such as the Rational*® Unified Process, a system is broken down into smaller components for the purpose of construction and deployment (Krutchen, 2000). While this is an effective risk-mitigation strategy, testing is often still seen as a post-construction activity[2] even though it occurs multiple times during a given project's lifecycle.

Testers are frequently considered to be "super users" of a system and in these cases, the overall quality of the system depends tenuously on their skills of discovery. Testers are often expected to take requirements of varying quality and extrapolate them into scripted test scenarios. These scenarios are documented as a narrative and are typically executed manually. In some cases, execution of these cases can be automated but in traditional practice, this is late-stage automation.

The end stages of traditional projects end up being heavy negotiation phases in which the development team and sponsors pour over the set of identified defects and unmet requirements. Such lists are often extensive as they cover a broad range of defects across the range of functionality implemented in the system. Furthermore, the defects are not limited to integration- and deployment-level defects (which are more understandable at that late integration phase), but instead include many unit-level defects in base functionality. While these situations are certainly symptomatic of other methodological concerns that many agile practices address, TDD practices have significantly limited the extent of such discussions in the author's experience.

Test-driven development often involves automation, but test-driven development is more than just the act of test automation. As we will see, there are a number of testing automation practices that are not, strictly speaking, test-driven practices—for example, using automated acceptance tools such as Fitnesse[3] to write acceptance scripts after implementation.

BACKGROUND

The perspectives on TDD presented in this chapter are based on the author's experience of applying TDD to four software projects in an industrial setting. These projects were implemented in the corporate IT organization at a Fortune 250 medical technology company. These projects were all browser-based Web applications.

The Projects

Project A was an information portal targeted at healthcare professionals who are customers of the company. Users complete a self-registration process, indicate their information preferences, and are required to login to view the site. The site then presents personalized navigation and content to users based on the user's professional specialties and geographic location. Content managers have additional access to post content to the site. This project consists of approximately 51,000 lines of code (LOC) and was developed by a team of six developers and an application architect.

Project B was an information portal targeted at internal field employees of the company. Users login to view the site and information is personalized automatically based on job role information in other corporate systems. Content is personalized to users based on job levels and geographic location. Content managers have additional access to post content to the site. This project consists of approximately 16,000 lines of code (LOC) and was developed by a team of three developers and an application architect.

Project C was a tool to assist internal field employees in generating pricing quotes for customers. Users have role-based access to different customer accounts and need to search available products to build a list of products for a quote. This system uses remote messaging to interface with backend enterprise systems that provided price information. This project consists of approximately 8,000 lines of code and was developed by two developers and an application architect.

Project D was a tool used by customer service agents to track customer requests related to insurance reimbursement. Agents interact with both customers and internal field employees when creating requests and are able to dynamically generate reports to respond to answer status requests and report performance to upper management. This project consists of approximately 16,000 lines of code and was developed by two developers and an application architect.

The People

The staff on all projects consisted of both employees and contract developers with at least four years of Java development experience. Developers had varying levels of experience with the specific development tools and frameworks used to build the applications, but were able to be productive with those technologies in a short time. However, only three of the developers had any prior experience writing automated programmer tests. The remaining developers had varying levels of informal exposure to unit testing, such as reading articles in trade publications or books. No developers had any formal training in TDD before or during these projects.

The Approach

Developers used Eclipse, JUnit, and Ant in an individual development environment. Test coverage was focused on interfaces that were integration points among subsystems in the applications. Developers were expected to write their own test classes and run them frequently through Eclipse during their regular development activity. As part of the standard development environment, Ant scripts were provided that run the entire suite of tests. The entire suite was run prior to deploying builds to a testing environment, and in some cases continuous integrations ran the test suite on a more frequent and regular basis.

LESSONS LEARNED AND BEST PRACTICES

Overall, the development teams involved in the aforementioned projects decided that TDD practices were a beneficial addition to the projects. Developers generally felt that using TDD made them feel more confident that their code met requirements and would not require re-work. Teams determined that the following lessons learned and best practices contributed to those feelings.

Tests Really do Need to be "Easily" Executed

The notion of tests being subjectively "easy" to execute was introduced earlier as a defining characteristic of TDD. But team members definitely felt that they were more inclined to execute tests if they ran simply (automated test fixtures and test setup) and rapidly. The longer tests took to run, the more frequently developers would lapse into "test after" practices—just working on the implementation and writing tests later.

Smaller is Better

Teams felt that writing "small" tests was more effective. Small in this sense meant that individual test methods were very focused around testing a small amount of functionality. Large test methods tended to require large amounts of fixture setup and resetting and became more difficult to read. Writing small test methods with descriptive names tended to be more self-documenting and easier to maintain over time.

The Only Implementable Requirement is a Testable Requirement

A self-respecting TDD purist would state that for every functional requirement there should be a test, as a matter of principle. That is to say if you are able to describe a functional business requirement that can possibly be implemented as working software, then you should also be able to describe a software test to verify that the requirement was in fact implemented.

While this view is a bit extreme, it suggests a useful thought experiment for even the more realistic TDD practitioner. A developer who is properly "test-infected" will receive a functional requirement or enhancement request and will first think about the test. Where in the test suite would such a requirement best be verified? What are pre- and post-conditions for the functional scenario? What are boundary and exception conditions, and what should fallback behavior be in those cases? Beck comments that there are just two simple rules involved—writing failing automated tests before writing implementations and removing duplication (Beck, 2003)[4].

It might not be possible or practical to first implement a test for every possible scenario of every identified requirement. But it is a sign of healthy test-infection when a developer thinks through the verification process before implementation, and implements tests for critical aspects of the implementation (most frequent, highest risk, etc.).

Base Test Coverage on Risk Assessment

The question of test coverage can be a bit controversial in the world of TDD. One might think that 100% test coverage is essential for true TDD. It's possible that in a situation of doing brand-new, greenfield development with full management support for TDD and quality that this is attainable. However, this is often not the case on development projects[5]. More often than not, TDD practices will be introduced during the maintenance of existing code, or there might not be full support in the organization for the practice of TDD. The answer to the coverage question is that you should have the *proper* amount of test coverage on a system, as well as that subjective term can be applied to a given set of circumstances.

Coverage can be measured objectively using test coverage tools that become part of the overall testing harness. Coverage tools are responsible for reporting which lines of production code are executed during the execution of testing code. These tools might be used during the initial development of testing code, but are more often used as part of integration activities to audit for proper test coverage. Summary coverage is reported in terms of the overall percentage of code that is executed during tests. Coverage tools also report specific lines of code that are not executed at all during tests, so that additional coverage can be added. Some coverage tools, such as Clover[6] for testing Java code, provide this detailed view in the form of browsable HTML pages with highlighting to allow developers to more easily navigate to untested code.

Our teams felt that 100% coverage was not practical. While the notion of 100% coverage is appealing standing on its own, it is not always practical or possible to achieve this metric due to various concerns (schedule demands or organizational support, for example). Our teams used informal processes of identifying key areas of risk, based on the following criteria to identify where test coverage was most critical:

- Requirements were compliance-oriented.
- Functionality required interfaces with other systems.
- Implementation required contributions from other teams.
- Functional components required more expensive integration testing iterations (due to resource availability, time-consuming manual testing processes, etc.).

Our teams identified requirements for testing during the high-level design process. While deciding questions of overall design direction, the teams would also identify which functional areas were to be the focus of TDD. This approach proved to be a successful method of gaining the advantages of TDD in critical areas, while mitigating concerns around time spent in test development.

Integrate Tests into a Continuous Integration Process

A perhaps less obvious gain when using test-driven development is that you will be encouraged to build your code more incrementally that you might otherwise do. Test-driven development follows a simple rhythm of defining an interface, writing a test against the interface, and then writing the implementation until the test passes. This rhythm lends itself very well to the notion of biting off small pieces of functionality at a time, and continually expanding the implementation. Developers do this with great confidence because their ever-growing test suite promises to alert them if any defects are introduced along the way.

Test-driven development leads to a certain "food pellet" effect in this regard—developers often favor smaller increments because they are more quickly able to receive the feedback on whether they have introduced defects. This fact makes test-driven development a very natural enabler for increased agility on projects.

As this behavior becomes more and more common during the daily individual development process, teams also are better positioned to implement a continuous integration practice. Continuous integration is an important agile development technique in which the process of integrating is automated. In continuous integration, a scheduled and automated process pulls the latest code from source control and then compiles, tests, packages, and deploys the code to an integration environment. This process is run extremely frequently—often multiple times during the work day. The goal of continuous integration is to alert the development team as soon as possible when there are conflicting changes and allow the team to respond sooner than in a more traditional process of integrating manually and less frequently. Some continuous integration frameworks, such as CruiseControl[7], include additional features such as email notification of build success or failure, and Web-based reporting of test results that greatly enhance a team's ability to respond to integration issues.

Teams following TDD practices are much better positioned for continuous integration because they will have a rich set of automated tests ready to be run. Running a full suite of automated tests is a valuable aspect of continuous integration, so TDD teams will be far ahead of teams that are writing tests later or not at all. Test-driven teams will arguably also have a more "well-gardened" set of tests that play well together, as the developers are in the habit of running tests continuously in their own development environments.

Continuous integration could be considered a form of test-driven packaging and deployment. The practice of writing tests first encourages developers to implement a design that is easily testable. Likewise, the practice of setting up a continuous integration environment encourages developers to implement a build and deployment process that is easily automated. Many more teams practice TDD than practice continuous integration and there is far more literature on the practice of TDD. But teams that practice TDD should consider continuous integration the next major step down the path of improve code quality.

Use Mocking to Address Components with Runtime Dependencies

A frequent issue in writing automated unit tests has to do with the issue of runtime dependencies. Production code often has more elaborate dependencies beyond the inputs to a given method, which are typically parts of the test fixture. Examples are dependencies on the availability of data sources, remote services, file system resources, and transports (such as an HTTP environment). The answer for addressing these additional dependencies is the use of mocking.

Mocking involves the use of objects that impersonate more complex runtime components. These mock objects are used just for the purpose of testing, and implement the minimum amount of behavior that is required by objects or components that are under test. There are almost always extra objects created in test scenario as part of the fixture. But the key difference between those objects and mock objects is that mock objects are stand-ins or façades for a more complex system of objects. The other test objects included in the fixture are typically valid object instances that consist of invented test data, but are valid object instances nonetheless.

The use of mock objects is invaluable when there is critical testing to be performed around objects with complex runtime dependencies. Some production-quality frameworks, such as the Spring framework[8], include mock objects that do most of the heavy lifting so the developer can focus more on writing the core tests. However mocking can also add significant overhead to test fixtures if the developer is solely responsible for creating the mock objects. In these cases, the developer should consider the return on effort for creating a more elaborate fixture, or should consider refactoring critical logic out of dependent objects to increase testability.

Teams Feel More Confident Making Changes

Our teams felt much more confident making significant functional changes when TDD was used. This feeling was especially evident when fixing defects discovered during the integration process or when implement enhancements to existing functionality. When developers could begin these tasks by executing a passing test suite and then writing new failing test (or modifying existing tests as appropriate so that they failed), they were much more confident that they were meeting the requirements they were given.

Teams Feel More Confident about the Design

TDD arguably encourages better design because the interface developer must first think of the client's needs. The test is the first client of the implementation and the developer is the first customer. The test is also a form of software documentation, which again increases overall quality. The test is a readable form of functional documentation that explicitly defines a functional software component. While this form of documentation might not adequately address higher business concerns or concepts around the functional requirements, it can explicitly demonstrate business rules in action. Tests might even been seen as a form of interactive documentation, as they can continually be added to when exploring boundary conditions or new requirements. In our experience, these factors contributed to a higher level of design satisfaction on teams.

TDD as a Refactoring Enabler

As a follow-up to the previous point, teams confirmed that TDD enabled much more productive refactoring. While this point has been discussed in literature (Beck, 1999), our teams certainly felt more confident in practice that refactoring was much easier to address when code was written using a TDD approach.

The code travels with its own suite for recertification, so changes can be more safely made later. The existence of ready-to-run tests enables maintenance developers to more confidently address enhancements and ensure system stability during later updates.

Loose Coupling and Component-Based Design Enable Testability

As we have discussed previously, the practice of TDD requires upfront thinking about design because the developer is writing a client before the implementation. Before arriving at that point, some basic principles of good application architecture and design can be applied in order to ensure that the overall system is constructed in a highly testable fashion.

Loose coupling of components is perhaps the most important principle that applies. Well-defined interfaces between subsystems greatly enhance the developer's ability to start small and

Figure 2. Example of subsystems in an enterprise Web application

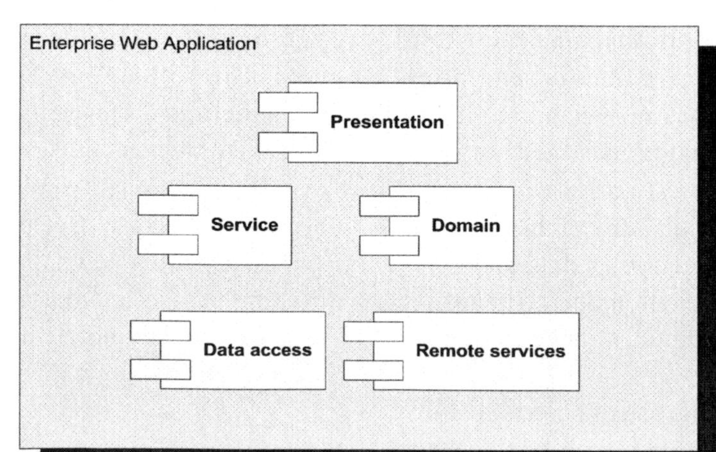

build a well-tested component for integration into a larger overall system. At the very least, development teams should agree on high-level subsystems or tiers and discuss interface needs between those components.

Figure 2 illustrates a typical subsystem architecture for a Web application with access to a local database as well as remote services.

A component-based architecture such as this enables a given subsystem to more easily be developed in isolation using TDD techniques. Implementation considerations of other subsystems can be safely disregarded so that the developer can focus on the core implementation of the given subsystem. We will later see how dependency mocking can assist with crossover areas and integration points.

Use of loosely coupled subsystems as a design technique has additional long-term benefits for defect isolation during maintenance phases. Future developers will have a much easier time isolating defects to a given subsystem, and will more likely be able to address defects by modifying the test suite and implementation for just a given subsystem.

New Roles for Architects and Testers During the Development Process

Test-driven development certainly empowers developers to become their own first testers. As important as this is, there are also impacts on the traditional roles of application architects and functional testers in the TDD world.

Architects must be able to ensure that the overall application design encourages testability. Architects must consider issues such as the following and determine their impacts on testability:

- Are there runtime dependencies that must be mocked for testing?
- Are subsystems and components properly isolated for testing discrete functionality?
- Is there a "culture of testing" on the team, so that test-driven development practices will be consistently followed?

Testers become a new type of subject-matter expert (SME) for developers on the topic of writing tests. Testers at the very least can help developers identify key rules for testing and corner cases that will drive refactoring. On larger teams or more critical systems, the testers might take an even more active role, at least at the outset. Testers might pair program with developers when writing interfaces and initial tests.

Integrate Testing Best Practices into Coding Best Practices

Writing tests involves writing additional code and test code should follow coding guidelines just as the implementation code does. But for testing code, teams should additionally define standards test-specific concerns. The following items are examples of topics to cover when defining testing best practices:

- Proper approaches for setting up test fixtures programmatically.
- Process for setting up environments—such as loading test data or frameworks for creating mock objects.
- Methods for assembling test cases into test suites.
- Scripts and processes for executing suites.

CHALLENGES IN TDD

While teams felt overall that TDD was an effective practice, it is not without its challenges. Many of these are perhaps the natural experiences of teams that are new to TDD, but there were real considerations in our experience.

Difficult to Get Started with TDD

Our experience was that many developers were interested in TDD, but had a hard time getting started writing their first test classes. Developers seemed confused about how to write assertions around code that did not yet exist or weren't quite sure where to begin writing tests within the context of a class design they had in mind. A mitigation strategy for this issue is to use mentoring or pairing to help developers get their first tests written, at least at a high level. A more experienced TDD developer can write initial rough tests, and coach the newer TDD developer through the process of expanding the tested scenarios.

There are sometimes organizational factors at work here. When there is a setting that encourages "just getting things done" or very rapid delivery, developers sometimes had a difficult time carving out time to write their first tests. Even if they have some tests written before implementation, writing tests was sometimes dropped lower on the priority list in favor of showing progress on functional code. If teams want to adopt TDD practices, our experience was that they needed to allow developers periods of lower productivity during early stages of adoption to acclimate themselves to the TDD approach.

Unique Challenges with TDD and Existing Code

There is a tremendous volume of existing production code that has no existing automated unit tests, and a significant portion of total developer time in an organization is spent maintaining this existing code. While it might not be as easy to introduce TDD practices on such code, it is not impossible. Our teams found that the quality of the overall design was the largest determining factor for how challenging it will to introduce TDD. Obviously, better designs with more isolation and loose coupling will lend themselves more easily to automated testing. But in any event, if code was not designed with automated testing in mind[9] it will likely require some amount of refactoring to become testable.

The process of introducing automated tests onto existing untested code should not be done wholesale, but should rather be done gradually in keeping with Agile principles of building to requirements and continuously delivering value (rather than embarking on extremely long term re-engineering efforts). The introduction of testing should be considered part of the defect and enhancement process. The required process change is that all defects will be verified with failing tests before being fixed, and passing tests will be written against affected areas before implementing test-first enhancements.

The developer must refactor the interface that requires testing in order to make it testable. This usually involves breaking dependencies and simplifying the interface so that an elaborate fixture is not required. A less desirable alternative is to implement mocking for the dependencies. This is less desirable because developer-implemented mocking is considered less valuable overall than refactoring the production code to improve code quality. And improving code quality is the primary goal of TDD in the first place!

Fowler (2004) described this process of gradual improvement over time while progress to an overall vision of reengineering as the "StranglerApplication." Fowler models his description after the Strangler vine, which symbiotically grows on top of a tree, but eventually overcomes and kills the tree. Likewise, introducing automated tests and TDD practices onto existing code should be seen as a Strangler vine that will eventually kill off the last bits of untestable code from a system.

Coverage is a Recurring Conversation

Our teams felt that 100% coverage was not a practical goal in our environment, so instead pursued partial coverage based on informal risk assessment. A consequence of this direction was that appropriate coverage was an ongoing conversation. Although the team valued the agile nature of determining test coverage from a business value standpoint, conversations on the topic were by definition very subjective rather than objective. Teams pursuing deliberate partial coverage should be prepared with strategies for bringing such conversations to closure so they don't consume valuable development time.

But are Tests Just More Code to Write? And isn't this Someone Else's Job?

For developers that are new to unit testing and test-driven development, writing test cases in code might just appear to be additional work. It certainly *looks* like additional work—its extra code that they didn't have to write before! Writing tests is an investment in the future. Some developers will immediately see the value of TDD; others will need to stick with it for awhile before realizing the value. Reluctant developers will gradually come around as they see tests in other areas of an application fail unexpectedly after making changes. The test suite will start to be less of a burden and more of a safety net.

Sometimes there is resistance to TDD not because it implies additional *tasks* but because it seems to imply additional *responsibility*. Traditional role delineations might have developers thinking that they are responsible for design and implementation and someone else is responsible for finding defects. The answer here lies in the fact that the developer's responsibility is to implement the requirements, and TDD should be seen as an advantageous method for ensuring those requirements are met.

Automation Alone does not Make for Test-Driven Development

It is worth emphasizing that there are many forms of automated testing, but automation alone does not make these test-driven development practices. The following are some typical automated testing practices, and while they certain contribute to overall system quality, they do not strictly fall under the umbrella of test-driven development.

User Interface Integration Testing

Full integration tests are sometimes run at the user interface level. Scripting is used to simulate user actions in the actual user interface, as a means of full regression testing of all involved system components. But in this case the system must be functionally complete for the tests to be executed, so this practice is certainly not driving development.

Load or Stress Testing

Load or stress testing is the practice of simulating heavy usage in a system. This is sometimes done to verify a non-functional requirement for system usage, such as number of concurrent users or transaction throughput. In other cases, this is a form of resiliency testing to determine the usage level at which some system components will fail or malfunction. This method of testing is almost exclusively done through automation and scripting. But here again the system must be functionally complete for the tests to be executed, so this practice is not driving development.

User Acceptance Testing

User acceptance tests are increasingly being executed through automation using tools such as Fitnesse. These tools allow users to specify the fixtures around tests, and then execute the tests themselves. There are ultimately coded test cases that execute behind the scenes, using the provided fixtures. But this type of testing is not considered TDD for the obvious reasons that it is automated, is not executed directly by the developer, and is executed after the implementation rather than before.

Objective Productivity and Quality Improvements are Debatable

Several studies (Erdogmus, Morisio, & Torchiano, 2005; George & Williams, 2003; Geras & Miller, 2004; Maximilien & Williams, 2003, Reifer, 2002) have explored the effects of using TDD vs. "test last" or "no test" development and assessed the impacts on developer productivity and product quality. Many studies did indicate a positive impact on product quality (Erdogmus et al., 2005; George et al., 2003; Geras et al., 2004; Maximilien et al., 2003), in one case even asserting that defects dropped by 50% (Maximilien et al., 2003), but researchers often had small sample sizes that they considered to be threats to validity. Changes in productivity were reported to be either neutral (Erdogmus et al., 2005) or negative (Erdogmus et al., 2005; George et al., 2003; Maximilien et al., 2003). The author does not have quantitative data around productivity on the four projects in question, so cannot comment objectively on productivity. While researchers have various explanations for these findings, there is not definitive, objective evidence to say that TDD makes developers or teams more productive.

Interestingly enough, a survey (Reifer, 2002) indicated an increase in productivity when using TDD; the survey was not coupled with any objective assessment. This fits with the authors experience—all developers using TDD practices on these projects shared that they felt more confident in their designs and code, and that they felt more productive. These less measurable benefits might be of value to readers from other perspectives such as maintaining staff morale and commitment to quality.

Overall, these findings and experiences make TDD a harder prospect to "sell up" in a large organization. Without more conclusive objective data, teams wanting to introduce TDD practices into their organizations need to base their cases on less tangible benefits, or do their own objective evaluations in their own environments to determine if TDD is well suited.

WHAT'S NEXT FOR TEST-DRIVEN DEVELOPMENT?: FUTURE TRENDS

The practice of TDD is well defined in current literature and it is gradually becoming a standard behavior in the development process, especially for teams that embrace extreme programming (XP) or agile techniques.

Integrations and plug-ins for many popular IDEs provide useful tools for test generation and execution. There are also many useful reporting tools to help teams digest summary test execution results when running a large test suite. However many of these tools currently help developers generate skeleton test suites that still require code to become useful tests. Improvements to these tools might come in the form of integration with code analysis tools that will suggest assertions to put inside the tests themselves. The ruby on rails development framework automatically generates stub test classes for declared implementation classes independent of an IDE integration—this is considered a core service of the framework[10]. This concept of integrating tests and implementation will likely become more common in development frameworks as the practice of TDD increases.

Core language enhancements will certainly influence how TDD is practiced, with features such as assertions, annotations, and other declarative programming[11] techniques playing a major role. Many current TDD practitioners believe that tests should be kept as focused as possible, and that tests with elaborate fixtures should really be broken apart into smaller tests. This will probably not be the case as more and more languages include the above features. These features can be used to perform equivalent tests on production code during execution, and therefore will encourage developers to skip writing tests at a very fine-grained level and instead focus on writing higher level integration tests.

CONCLUSION

There are many ways that TDD can take root in an organization, team, or individual. At the organization level, it can be the top-down promotion of the idea that quality is everyone's responsibility, not just the designated Quality Assurance team. At the team level, it can be a means of saving cost and time in the development cycle by ensuring that the expensive cycles of human-performed integration tests are used wisely for performing true integration tests rather than discovering defects that should be caught earlier. At the individual level, it can be the realization that writing tests first can prevent tedious re-coding later and can be far easier than breakpoint debugging for ensuring that code meets expectations.

There are many benefits to adopting TDD, but it is not without challenges. Learning new behaviors, making additional decisions around coding standards, and deciding on test coverage are just a few challenges teams will confront when adopting TDD. Awareness of these challenges will help teams address them upfront, and might also serve as an example for teams to continually evaluate what challenges they face in their own environment. TDD is not a panacea for all testing concerns in developing software, but it can certainly contribute to a team's commitment to improving the quality of their software.

REFERENCES

Beck, K. (1998). *Kent Beck's guide to better smalltalk*. Cambridge, UK: Cambridge University Press.

Beck, K. (2002). *Test-driven development: By example*. Boston: Addison Wesley.

Erdogmus, H., Morisio, M., & Torchiano, M. (2005). On the effectiveness of the test-first approach to programming. *IEEE Transactions on Software Engineering, 31*(3), 226-237.

Fowler, M. (2004). *StranglerApplication blog entry on martinfowler.com*. Retrieved January 30, 2006, from http://www.martinfowler.com/bliki/StranglerApplication.html

George, B., & Williams, L. (2003). A structured experiment of test-driven development. *Information and Software Technology, 46*, 337-342.

Geras, A., Smith, M., & Miller, J. (2004). *A prototype empirical evaluation of test driven development*. Paper presented at 10th International Symposium on Software Metrics (METRICS '04), Chicago.

Krutchen, P. (2000). *The rational unified process: An introduction*. Boston: Addison-Wesley Professional.

Maximilien, E. M., & Williams, L. (2003). *Assessing test-driven development at IBM*. Presented at the 25th International Conference on Software Engineering, Portland, OR.

Rainsberger, J. B., & Stirling, S. (2005). *JUnit recipes*. Greenwich, CT: Manning Publications, Co.

Reifer, D. (2002). How good are agile methods? *IEEE Software*, 16-18, July/August.

ADDITIONAL RESOURCES

JUnit: http://www.junit.org
CPPUnit: http://cppunit.sourceforge.net/
PyUnit: http://pyunit.sourceforge.net/
NUnit: http://www.nunit.org/
DBUnit: http://dbunit.sourceforge.net/
JWebUnit: http://jwebunit.sourceforge.net/
HttpUnit: http://httpunit.sourceforge.net/
HtmlUnit: http://htmlunit.sourceforge.net/
CruiseControl:
http://cruisecontrol.sourceforge.net/
FitNesse: http://fitnesse.org
Clover: http://www.cenqua.com/clover/

ENDNOTES

* Rational® Unified Process is a registered trademark of IBM Corporation.

1 For the sake of simplicity, all comments will refer to a single developer running tests. In practice, TDD is a very common practice combined with pair programming in the extreme programming (XP) methodology. Pairing is another valuable but sometimes controversial topic altogether.

2 RUP purists will note here that the RUP is considered a use-case driven process, and use cases are seen as predecessors to test cases. From a process standpoint, the requirements are perhaps more readily testable or lend themselves to testability, but the process itself is not test-driven in the sense that it does not specify the construction of testing components before functional components. TDD can nonetheless be applied to a process based on RUP.

3 See http://fitnesse.org/ for additional information.

4 Beck eloquently and concisely states that TDD "encourages simple designs and test suites that inspire confidence" (Beck, 2003).

5 The authors of JUnit Recipes (Rainsberger, Stirling, 2005) note that they almost always need to deal with legacy code. The value of applying TDD to existing code should not be underestimated.

6 See http://www.cenqua.com/clover/ for additional information.

7 See http://cruisecontrol.sourceforge.net/ for additional information.

8 Spring provides very useful mocks for HTTP, JNDI, and JDBC components. See http://www.springframework.org for additional information.

9 Automated testing is a non-functional requirement, and in this case the developer is really enhancing the code to meet new requirements.

10 See API documentation at http://www.rubyonrails.org/ for additional information. Test stub generation is built into the "scripts/generate model" command.

11 Declarative programming is the practice of stating end conditions as fact declarations, and allowing an underlying framework to take care of procedural concerns to arrive at the declared state. Procedural programming, on the other hand, is defined by algorithms that are responsible for processing and state transition.

Chapter XII
Quality Improvements from using Agile Development Methods:
Lessons Learned

Beatrice Miao Hwong
SIEMENS, USA

Gilberto Matos
SIEMENS, USA

Monica McKenna
SIEMENS, USA

Christopher Nelson
SIEMENS, USA

Gergana Nikolova
SIEMENS, USA

Arnold Rudorfer
SIEMENS, USA

Xiping Song
SIEMENS, USA

Grace Yuan Tai
SIEMENS, USA

Rajanikanth Tanikella
SIEMENS, USA

Bradley Wehrwein
SIEMENS, USA

ABSTRACT

In the past few years, Siemens has gained considerable experience using agile processes with several projects of varying size, duration, and complexity. We have observed an emerging pattern of quality assurance goals and practices across these experiences. We will provide background information on the various projects upon which our experiences are based, as well as on the agile processes used for them. Following the brief discussion of goals and practices, this chapter will present the lessons learned from the successes and failures in practicing quality assurance in agile projects. We aim at informing fellow agile developers and researchers about our methods for achieving quality goals, as well as providing an understanding of the current state of quality assurance in agile practices.

INTRODUCTION

Since the declaration of the agile manifesto (Beck et al., 2001) in February 2001, agile software development methods have enjoyed a proliferation leading to the spawning of variants and a proselytizing of agile methods as silver bullets (Brooks, 1987). Many Siemens organizations are turning to agile methods in order to shorten product development timelines. Siemens Corporate Research (SCR), the R&D center for Siemens in the U.S., has even formed its own agile development group. This chapter will discuss project experiences that this SCR group has been involved in to show how quality is approached in agile development.

The background section that follows will provide an overview of seven projects in which in-house agile processes were used. Next, there will be a discussion of common quality assurance (QA) goals and practices amongst these projects. This discussion will lead up to the section on lessons that we have learned so far and then conclusions for improving QA in future agile projects.

BACKGROUND

Within this section, we introduce two Siemens in-house agile processes, along with seven projects in which they were employed. The first process, named S-RaP (an acronym for **S**iemens **Ra**pid **P**rototyping), is a UI (user interface)-centered workflow-oriented approach that targets primarily the exploration of complex business requirements. The second process, entitled UPXS, is a combination of traditional and agile practices (**U**nified **P**rocess (Jacobson, Booch, & Rumbaugh, 1999), e**X**treme **P**rogramming (Beck, 1999), and **S**crum (Schwaber & Beedle, 2001)) that aims to

Figure 1. S-RaP process model (Nelson & Kim, 2004) (image by Kathleen Datta)

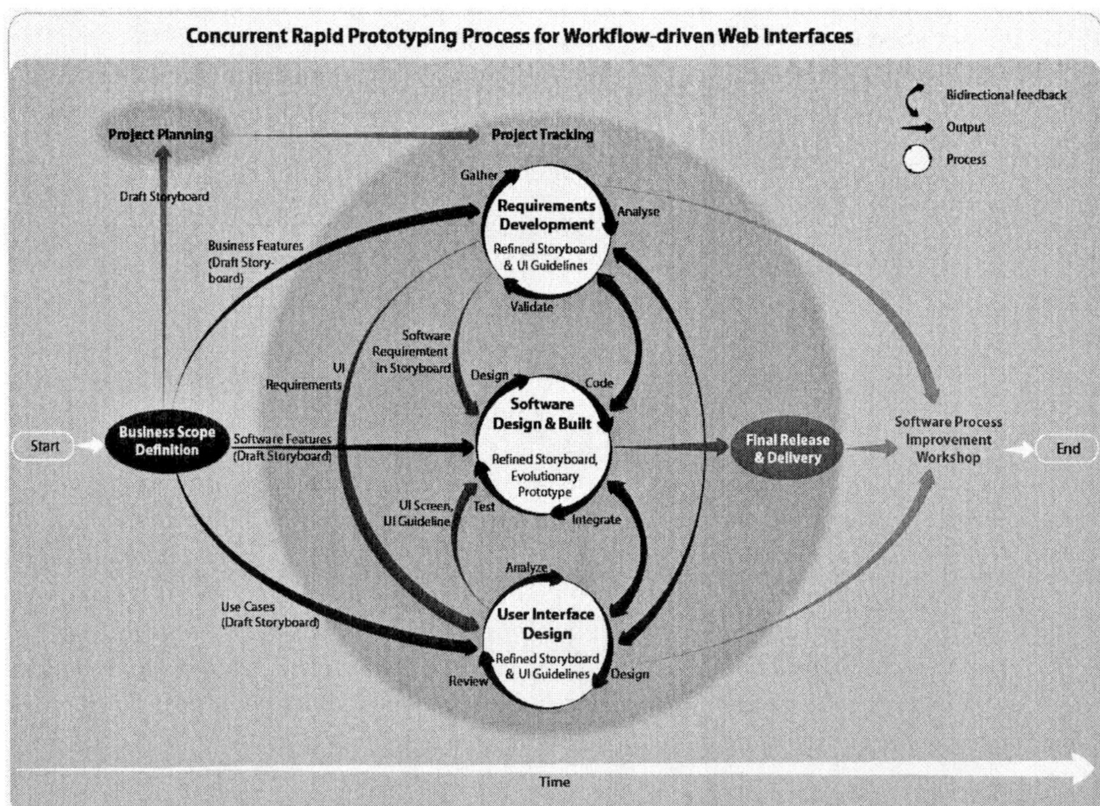

support full-blown product development (even product-lines).

SIEMENS AGILE PROCESSES

S-RaP

The S-RaP (Gunaratne, Hwong, Nelson, & Rudorfer, 2004; Hwong, Laurance, Rudorfer, & Song, 2004; Nelson & Kim, 2004; Song, Matos, Hwong, Rudorfer, & Nelson, 2004; Song, Matos, Hwong, Rudorfer, & Nelson, 2005; Tai, 2005) process evolved to provide rapid prototyping solutions for Siemens customers. An S-RaP project starts with the identification and prioritization of a set of business features by the customer and proceeds according to the "time-boxing" technique (McConnell, 1996). The features with higher priority are developed first and then the remaining features are addressed as time permits. An illustration of the iteration, concurrency, and coupling of the S-RaP process is presented in Figure 1.

S-RaP development is concentrated around two key artifacts:

Storyboard

The features planned for development are organized into workflows. These workflows are pictured and described in the context of stories within the Storyboard. An example of one such Storyboard appears as Figure 2.

The Storyboard is the requirements and testing specification for the developers and a means to establish and communicate a common product vision among all stakeholders. It is evolved iteratively, integrating customers and domain experts to work together with the development team toward the identification of new, and clarification of existing, requirements. Due to the limitation of static content, the Storyboard alone might not be capable of clarifying the full semantics of the user interface and the workflow interactions. This problem is solved by the availability of an interactive prototype early on.

Figure 2. Sample screenshot of Storyboard (Song et al., 2005)

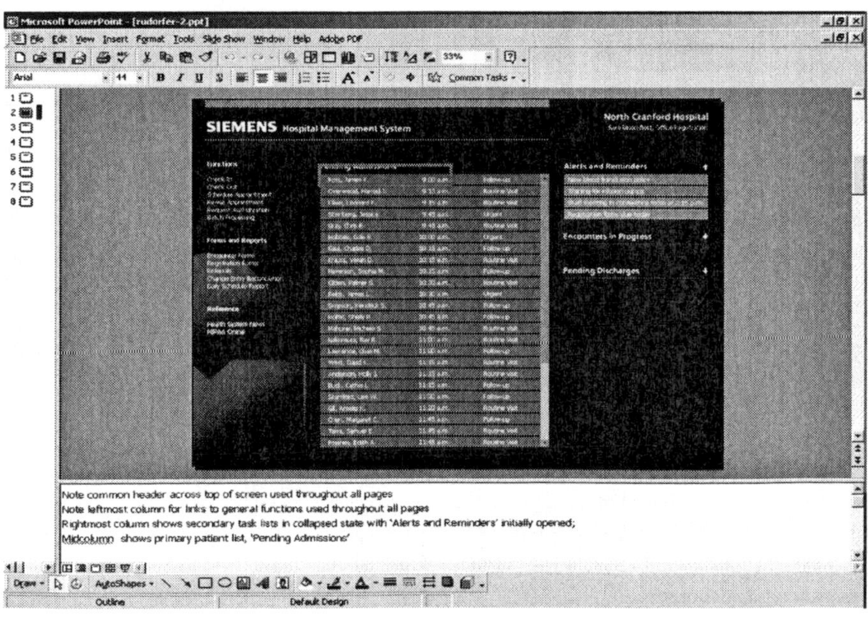

Prototype

During development, the software prototype provides the customer with a working representation of the final deliverable at an early stage. This gives the customer a hands-on testing experience with what has been developed, which helps to validate existing ideas and oftentimes generates new ones. The developers start the prototyping and testing activities as soon as some workflow features become stable. They select the development tasks autonomously and provide the team leads with coarse time-effort estimation.

Note that both activities—Storyboarding and prototyping—run quasi-concurrently. While the UI designers and requirements engineers model the UI and its functionality in the Storyboard, the developers implement the related features in the prototype. The iterative, quasi-concurrent evolution of these two artifacts allows dynamic consideration and integration of the customer needs (requirements) in the process and supports the delivery of a product which meets these needs.

UPXS

UPXS (Pichler, 2006; Smith & Pichler, 2005) is an agile development methodology that combines principles from the unified software development process (UP), XP, and Scrum. Developed for a high-profile Siemens project, the process was designed to address the needs of a large distributed agile project. With a foundation of Scrum's team structure and activities, UPXS adds the project timeline model and phases of UP, along with iteration and task planning and development practices from XP.

Similar to S-RaP, UPXS is executed in time-boxed iterations of 10 to 20 working days. Iterations begin with the selection and prioritization of a subset of features to implement from the product backlog. The initial backlog is created by the product owner to establish and prioritize the full set of features in the final deliverable. Unlike in S-RaP, UPXS prescribes the creation and use of more traditional artifacts, though they are created iteratively and evolve throughout the project. These artifacts include use cases, requirements documents, and software architecture documents. Project management is performed using project and iteration backlog documents to maintain a prioritized record of remaining tasks. and burn-down charts are used to track sprint progress. Daily Scrum meetings allow synchronization of team members and the escalation of any blocking points for the Scrum team. Similarly, a daily Scrum of Scrums meeting facilitates communication between Scrum teams and gives management an ongoing awareness of project status. Additionally, the Scrum of Scrum aids in coordinating distributed teams.

An important motivation behind UPXS is to provide greater predictability and risk management to project leaders. For this reason, UPXS takes daily and weekly progress as an input for feature selection and prioritization for coming iterations. Although the manner is not as hands-on as in S-RaP, product walkthroughs at the end of each iteration keep the product owner well informed of the current state and quality of the evolving software deliverables. With the four major phases into which iterations are grouped—Inception, Elaboration, Construction, and Transition—there are also milestone checkpoints for project management to evaluate the overall progress of the project.

Agile Projects

Over the past few years, the processes previously described were employed in a number of different projects. The experiences mentioned in later sections are taken from the seven projects we present in this section. Table 1 outlines the specific agile characteristics of these projects and is followed by more in-depth descriptions.

Table 1. Project characteristics

	Project A1 Medical Marketing Prototype	Project A2 Medical Marketing Prototype	Project B Medical Requirements Elicitation Prototype	Project C Medical Product	Project D Communications Platform	Project E Communications Product	Project F Building Technologies Prototype
Process	S-RaP	S-RaP	S-RaP	S-RaP	UPXS	UPXS	S-RaP
Iterations (Length, #)	1 to 2 weeks, 4 - 20	UI: 1 to 2 wks DB: 6 weeks	5 days	2 to 3 wks, ~30	4 weeks, 16	3 to 4 weeks, 8	2 to 3 wks, 5
Sites	2	5	2	5	6	4	2
Team members	3 to 14	20 to 35	7	10	40 to 100	100	13
Lines of code	576,525	170,000	21,000	58,000	500,000+	1,500,000+	6,500
Duration	1 year	1 year	12 weeks	1.5 years	1.5 years	28 weeks	24 weeks

Project A1: Medical Marketing Prototype

Project A1 was an S-RaP project focused on building upon an inherited prototype to produce a marketing demo of new advanced features and a modernized look and feel for an emerging software product. The application ran in a Web browser and used simple HTML and JavaScript technologies. Deadlines were fixed but features were often negotiated during the actual development. Since the product was intended for marketing purposes, the customer needed a reliable solution with a high fidelity UI.

Project A2: Medical Marketing Prototype

When the customer in Project A1 desired an advanced set of features that could not be easily done with its existing architecture, Project A2 was born. Project A2 was also an S-RaP project, but unlike its parent, A2 produced a prototype that consisted of a 3-tier architecture with an HTML-based UI, a server component, and an underlying database layer. The customer had decided to incorporate more workflows and an underlying database layer to enable dynamic configuration of workflows and application data. Thus, the customer desired a more maintainable and adaptable solution that was still reliable and also had a high fidelity UI.

Project B: Medical Requirements Elicitation Prototype

Project B is a smaller S-RaP project that produced a prototype starting from a vague statement of customer needs. There was no existing software upon which to base this prototype. Thus, it was critical to the project's success for requirements to be elicited and refined efficiently. Developers participated in defining the requirements by providing suggestions on how to model the interaction features that were not yet fully specified, which contributed to their practical viability. The customer's main desire was to elicit, specify, and verify requirements for features of a new product. The development of a high-fidelity UI prototype delivered not only the clarified requirements but also a demonstration tool that could be used for collecting further feedback from prospective clients.

Project C: Medical Product

Project C is another S-RaP project that produced a small 3-tier product. Although S-RaP was originally developed for prototyping purposes, this project showed that the prototyping process could also support the development of a finished commercial product. Project C lasted 1.5 years and began as a prototype used to explore UI, ease-of-use, and performance issues, after which development began on a second prototype that evolved into the final deliverable. The practice of S-RaP principles helped us collect useful information that influenced the requirements and design of the final solution. In terms of quality goals, the focus was initially on high security, so as not to compromise personal data, as well as a highly attractive and easy-to-use UI. Once these goals were met, the focus shifted towards performance.

Project D: Communications Platform

Project D is a UPXS project that began with a mostly centralized co-located team and has expanded into a worldwide-distributed project to develop a groundbreaking platform upon which future communications applications will run. In true agile manner, this n-tier service-oriented framework continues to be developed in parallel with Project E, supporting its ongoing needs. With a final deliverable that has no UI, this project had quality goals focused on achieving high reliability, portability, maintainability, security, and performance.

Project E: Communications Product

Project E is a UPXS project with a large number of distributed teams working on a product that will replace several legacy applications. The Web-based application interfaces with databases and communication hardware, and runs on the framework that is simultaneously evolving in Project D. The main goal of this project is to produce a high quality UI that integrates functionality from the legacy applications in a performance-enhancing and highly intuitive way. At the same time, this product serves as a source of requirements elicitation and refinement for Project D.

Project F: Building Technologies Requirements Elicitation Prototype

Project F is the smallest S-RaP project yet, which aimed to elicit, refine, and mature the requirements for a next generation product. This prototype client-server Web application was intended to serve as a form of requirements specification for the development of a new product that would integrate functionality from and replace several legacy applications. Since the functionality stemmed from existing applications, the focus was more on developing an innovative, high-fidelity UI that would still be intuitive and useful to customers of the existing applications.

QUALITY ASSURANCE: GOALS AND PRACTICES

"QA (for agile methods) is looking at the same deliverables (as with plan-driven methods). But the process used to create the deliverables does affect how QA works" (McBreen, 2002). Our experiences have shown us that the cycle of customer involvement—constant re-estimation, and constant reprioritization of scope and features—is an inherent mechanism of agile methods that leads to high software quality.

Common Quality Goals and Practices

Although each of our projects focused on their own set of quality goals, there were several common goals that were important to all of them. The following outlines these goals and the QA practices that we applied to achieve them in one or more of our projects.

Goal 1: *The final deliverable should exhibit a high degree of correctness of implementation.*

Incorporating testing practices as soon as possible and on multiple levels (unit, integration, and system) was a technique we used to ensure correctness of implementation. For example, in our UPXS projects, developers wrote unit tests, which were continuously run as part of the build process. Simultaneously, a test team developed and ran integration and system tests.

A similar measure of correctness in our S-RaP projects was achieved by acceptance testing, which directly involved the customer. Unlike traditional acceptance testing that starts once the final product is delivered, acceptance testing in S-RaP was a constant and continuous process. With the end of each iteration, the customer could execute acceptance tests on the part of the system that was delivered.

Collective ownership was another technique used to help ensure correctness of implementation by encouraging developers to look at and improve each other's code. This form of peer review, like the XP practice of pair programming, increases knowledge sharing amongst developers, and can expose deficiencies in the implementation. In our UPXS projects, this resulted in explicit communication of improved or best practices for specific development tasks on several occasions.

Goal 2: *The final deliverable is well suited to the expressed needs of the customer.*

One of the key practices we have used to achieve this goal is to incorporate the customer in planning and verification activities throughout the project. Without beginning-to-end involvement of the customer in a project, it is possible that the resulting software is unsuitable for the customer's needs despite meeting all the customer's requirements. These needs should be captured in the requirements; however, a situation of this nature can arise when there is a mismatch, miscommunication, or omission of key project requirements.

By ensuring that the customer is involved in every aspect of the project, from planning of requested features and definition of requirements to continued verification of software deliverables, misconceptions are reduced, and the result is a product more in line with the expressed customer needs.

Goal 3: *The final deliverable is easy-to-understand, easy-to-learn, and easy-to-use.*

Since the customer is the key stakeholder who decides if the final deliverable is attractive, easy-to-understand, easy-to-learn, and easy-to-use, our technique for achieving this goal focuses on early and frequent customer involvement. From very early stages, we involve the customer in hands-on walkthroughs of the working software. The early feedback from this method drives development toward achieving this goal right from the outset. In our S-RaP projects, the Storyboard drove the customer feedback cycle.

Goal 4: *At any stage of development, code is easily analyzable, modifiable, and extensible.*

Throughout the development of the software deliverable, it is necessary to accommodate constantly changing requirements without requiring significant rework, as well as embrace requirements for new features. Several practices we have successfully used to maintain code simplicity include keeping designs simple and refactoring, as discussed in Lesson 9, which can be found in the Lessons Learned section.

Additionally, since later modifications to software are often necessary, we practiced test-driven development to ensure that these changes are smoothly integrated. Unit tests are particularly useful for ensuring early and continuous regression testing. The availability of a large number of tests of sufficient breadth and depth of functionality, in combination with high code coverage, is a significant contributor to achieving modifiability, because it provides developers with the confidence to make changes knowing that they have a suite of unit tests as a safety net.

LESSONS LEARNED

From the seven different project experiences that this chapter draws upon, each with a different set of goals, we learned many lessons about the application of our agile methodologies and practices. This section generalizes these lessons and presents them in their order of significance.

Lesson 1: *Use "living" documents whose lives are determined by the project's needs.*

Our experience generally confirms the viability of development with a reduced emphasis on producing detailed documentation, which is one of the values of the agile manifesto (Beck et al., 2001). In both the prototyping and product-oriented projects, we saw that the most important forms of documentation were the "living" documents, which informally captured multiple views of a specific problem and addressed different stakeholders' concerns. Such living documents were available to all team members for editing, thereby constituting a collaboration and communication medium. Most of the targeted documents that dealt with a specific architectural or product-related issue had a short shelf life and became stale and out-of-sync with the evolving code base. We have found that the best way of representing and discussing requirements is in a form that is very close to both the users' intuitive understanding of needs, and to the developers' understanding of context and presentation for the solution. Such a collaborative medium has been a significant aid to requirements gathering and forming an understanding and consensus between team members charged with different roles and/or bringing different skills and perspectives to the table (e.g., domain knowledge vs. UI and interaction design vs. implementation skill.

In our S-RaP projects, we used Microsoft Office PowerPoint as a tool to present the sequence that illustrated a feature of interest, earlier introduced as the Storyboard. All stakeholders tied to a Storyboard could add screen wireframes, or screenshots from the evolving product or related applications, and then use the presentation editing facilities of PowerPoint to annotate the image with the desired product functionality. Similarly, the notes section of the PowerPoint presentation was used for the textual description of the specific interactions and data issues related to each slide's illustration. Though not without its shortcomings, we found the Storyboard to be effective in providing information that is useful to software developers, while preserving the immediate intuitive nature of a graphically aided interaction sequence.

Documents that succinctly capture the most relevant requirement details from stakeholders, like the Storyboard, are very useful from the development team standpoint. When formal requirements engineering (RE) processes are used, the result can be very detailed, inter-correlated documents that still fail to present the global view that connects the requirements to their purpose. From our project experiences, it has made a big difference when the development team pushes for the use of consolidated, interactive communication formats that embody specific input from all stakeholders in a project (i.e., UI, Usability, RE, etc.). Though its contents may prove redundant with the artifacts of other contributors, it will enable more efficient progress—the results of which should eventually be captured more formally.

We have also experienced projects without living documents. Often times the lack of these documents, also referred to as "boundary objects"

(Gunaratne et al., 2004), leads to frustration within teams, and miscommunications between teams. Without these documents, team members often struggled to find correct and up-to-date information pertinent to their tasks. These documents provide a necessary medium for communication amongst and across teams to ensure a common understanding.

Lesson 2: *Development needs to be proactive with the customer by providing solution alternates in a form easily grasped by the customer.*

The agile development team is responsible for ensuring the project progress, and that implies that they must push for the identification and implementation of solutions for any problems that the product needs to address. The customer or product owner is responsible for making decisions on project direction, and doing it in a timely manner in order to allow the developers to proceed on the priority areas. The general loop of decision requests starts with the development group, which identifies an issue that needs clarification, and then shifts to the product owner, who may need to ask other stakeholders for more information before a decision can be made. There are two specific approaches to improve the efficiency of the decision-making process:

- Imposing decision deadlines on the product owner, and
- Proactively providing the product owner with a selection of viable solutions (instead of just asking a general question about the issue).

Short decision deadlines are a simple way of tightly integrating the product owner into the time schedule constraints of the development team. The proactive approach of partially elaborating promising solutions before submitting them to the product owner for a decision plays a much more important role in speeding up the innovation and solution cycle between the developers and their customers. It is our experience that a product owner or domain expert, faced with an open-ended question on what they would like to see in a given feature, is more likely to make a detailed decision if they are provided with examples that they can use to reason about their preferred solution. Since the domain expert is commonly a critical resource, providing them with some exploratory results related to the decisions under consideration helps to maximize the impact of their involvement. We have seen very good results from doing some Storyboarding of viable solution alternatives and adding that information to the decision request presented to the client.

Lesson 3: *Inexpert team members can be agile; however, the learning curve will be significant for those who lack technical expertise.*

One common complaint about agile development is that in most cases their success depends on having teams of experts (Turner & Boehm, 2003), both in the technical and the application domains. On the contrary, in our agile experiences, we have seen team members with less-than-expert domain knowledge quickly adapt to developing in an agile environment. Most of our projects included a significant number of team members who had minimal experience and knowledge in agile development and the project's domain. In Projects A1 and A2, developers who had a mid-level proficiency in the selected software technology but no domain knowledge were able to start contributing within a couple days.

This is not to say that no technical expertise or domain knowledge is required for new members to be integrated into an agile development process. Our experience has been that members with below-level technical skills will face a steep learning curve that is magnified by the nature of agile development. The quick evolution of the developed code means that less experienced developers cannot benefit from any stable code infrastructure to use as a reference. For instance, one such developer in Project E had good domain knowledge but below-average technical skills, and this individual was never able to reach a level of parity with the other developers. On the other hand, pairing new members with experts decreased the learning time for new members. Daily meetings, as used in Project B, helped in making performance issues more transparent.

It has also been observed that new members need to adjust to the unfamiliar demands of an agile process. For example, developers' code changes have to be completed, integrated, and tested in hours or days, instead of weeks. Developers also need to be able to shrug off the fact that a new decision, project constraint, or requirement could suddenly make their envisioned designs or previous work obsolete. Additionally, as customer demands change during product development, new code segments and interfaces may appear that need to be learned quickly. Code refactoring can change an interface that was finally agreed upon last week to something completely different this week. Developers also have to learn to fix or work around broken builds to avoid being blocked—unable to continue with development. These are just a subset of the demands we have seen placed on developers in an agile environment.

Although most developers with adequate technical experience found the agile process intuitive, we have also seen technically skilled team members, unaccustomed to agile environments, having a difficult time adjusting. On more than one occasion, especially in Project E, we experienced team members who could not adjust to the more free-form nature of agile processes. Scrum masters or other leaders had to follow a more prescriptive approach with these team members. Small, detailed tasks were often specified, and specific milestones, within iterations, were set for these tasks to be completed and then reviewed.

Lesson 4: *Agile methodologies must be practiced with a culture of proactive communication to allow new members to acclimate quickly.*

With practices such as self-documenting code (Martin, 2003; McConnell, 2004) and just-enough documentation (Beck, 1999; Turner & Boehm, 2003), the successful execution of agile development is dependent on team members receiving information through electronic correspondence, informal discussions, meetings, or the code itself. One side effect of using minimal documentation is that there is no explicit source explaining how people in the development team know what they know. "Agile methods rely heavily on communication through tacit, interpersonal knowledge" (Turner & Boehm, 2003). From the standpoint of new team members, it is difficult to identify the correct sources for necessary information. In S-RaP, we address this problem with the Storyboard, but in general, we found that it is important for all team members to proactively communicate with new members to help them transition into the project.

Lesson 5: *Agile development needs agile project planning (Song et al., 2004).*

Project planning should be the most agile part of agile development, starting with a coarse scoping and chunking of deliverables, and then refining the estimates as the progress provides more data. In agile development, a great emphasis is placed on achieving timely delivery against tightly scheduled milestones. Unfortunately, estimates for project deadlines may often be derived from only the initial understanding of requirements, as was our experience in Project E. Since this set of requirements is expected to be incomplete and vague, such estimates will often be unreliable. In the case of Project E, unrealistic deadlines set in the early stages were perceived as a source of problems throughout the project due to continuously reported delays. In this particular project, we were able to get back on track with the preset milestones through task reprioritization and scope adjustment.

Lesson 6: *To achieve high customer satisfaction in agile development, collecting novice user feedback is just as important as regular customer feedback.*

The most visible strength of agile development is in being able to achieve high customer satisfaction. Customers that are highly involved in the definition and refinement of project goals tend to be happier with the final result. An interesting effect of the constant involvement of customer representatives (i.e., product owners or domain experts) is that their expectations are affected. In Project B we realized this could also have a negative impact when a separate customer representative was presented with the final prototype and found it not-at-all intuitive, even though the customer representative who had been involved with the project had been very satisfied with the intuitiveness of the UI. Thus, the perception of intuitive quality of the product can be quite different for a first-time user. Novice user feedback would not only have helped in discovering this usability issue, but also in estimating the training needs and detecting any embedded idiosyncrasies that detract from the product's overall quality.

Lesson 7: *Collocation, when necessary, is best practiced within small teams (Song et al., 2004).*

Although collocation is a key practice of many agile development methodologies that foster informal, lightweight communication and leads to quick effective problem solving, it is not critical for all project teams to be collocated. Full-time collocation or even physical or geographic proximity is not required for teams working on well-partitioned vertical slices. In Project E, for example, after one of the collocated development teams moved to Greece, a vertical slice was assigned to this team, and this move caused virtually no disruption in the project schedule. For projects A1 and A2, we used instant messaging, teleconferencing, and online meetings to compensate for lack of collocation.

Depending on the size and scope of the project, our "small teams" consisted of 2 to 12 members. These smaller teams generally benefited from quick informal discussions with members working on similar tasks. However, with larger teams of more than 12, such as in Project E, this practice proved to be oftentimes more distracting than beneficial.

Lesson 8: *Decomposing project tasks to assign to different teams works best with vertical slices.*

Across our projects, we have seen multiple ways of decomposing the projects for concurrent development. Decomposition into vertical slices of functionality, where each sub-team was responsible for a UI segment and shared responsibility for its supporting layers, worked very well, provided that the sub-teams communicated about their work on common components. Continuous integration, nightly builds, and constant regression testing also helped to alleviate the headaches of integrating multiple vertical slices in projects D and E.

Although decomposition into horizontal layers worked well if the infrastructure layer had stable requirements that did not require refinement, it can also lead to more problems with synchronization. In Projects D and E, two simultaneously evolving projects, where the latter depended on the former, horizontal decomposition was used (in addition to vertical decomposition), and this raised complex issues of compatibility and synchronization between co-dependent iterative development activities. The communication that was needed in order to synchronize these activities was at the level of what is usually only available within a team, not between collaborating teams.

Another benefit of doing vertical decomposition of project functionality is that it allows partial decompositions and sharing of tasks between teams. The shared tasks encourage the sharing of developers across teams, and allow members of distinct teams to take over the communication and coordination responsibility for the specific shared components.

Lesson 9: *Where practical, postpone refactoring until several instances of the part under consideration (component, behavior, etc.) are implemented.*

Many of our projects were characterized by parts that were largely similar to each other. For example, Projects A1 and A2 included interactions that significantly resembled each other. For Project E, many UI implementation aspects varied only slightly in design. Such strong similarities, coupled with agile development's rapid nature and its emphasis on doing just enough for a specific delivery, lead quite naturally to the use of a copy-and-paste style of software development. We found that implementing several features independently in this manner accentuated the points of commonality, as well as the points of difference. This translated into implementation-level requirements that might not have otherwise been foreseen. It is precisely these requirements that provided the strongest guidance to refactoring.

While this approach has certain drawbacks, it is important to note that its negative effects are mostly indirect. For example, copy-and-paste leads to more maintenance work on the code or embedded documentation, but it generally does not lead to functional errors in and of itself. We have found that the opposite approach of trying to over-engineer reusable implementations too early tends to lead to both types of problems, functional failures and increased code development and maintenance cost.

That said, one caveat on this approach emerges from Project E: The act of refactoring copy-pasted portions of code is only manageable when the copy-pasted fragment has not proliferated too much in the code base. In Project E, the presentation-layer (Java Server Pages) was developed separately but concurrently by several developers in distributed teams to meet a specific visual layout/look-and-feel. However, even as these pages produced correct output, their internal document structures were different enough to create a maintenance nightmare without refactoring. The sheer number of these pages, coupled with the speed with which they were completed, amounted to a sizeable refactoring task; estimates for the additional effort were difficult for the development team to make.

The task of refactoring is a complex undertaking that is comprised of three separate subtasks: Recognizing an appropriate refactorable chunk, devising a refactoring solution, and adjusting/replacing the code that would be made obsolete by that solution. The point of this lesson targets the first of these subtasks to avoid premature over-engineering of such code bits—a lesson we consider quite important. The point of this caveat, however, is to warn against the potential bottleneck that the last subtask can become in agile development. If too much code is subject to refactoring, then the additional effort to adjust/replace this code will be difficult to estimate and can be unexpectedly substantial.

Finally, there are situations where the development team has far more control and internal knowledge, and where the early and proactive engineering of reusable and scalable solutions is mostly a positive approach. The overall software architecture for a product, for example, will usually be defined very early in the project, and these aspects tend to remain largely stable. Wherever the development team can identify improvements to the architecture or components which they consider useful from the development reuse standpoint, those are more likely to be stable since they often do not depend on the modifications of the explicit user interaction which may be requested by the customer.

Lesson 10: *A high level of customer satisfaction can still be achieved, even if the resulting deliverables do not entirely meet expectations.*

Due to the increased emphasis on customer communication and involvement in agile processes, we have found that the resulting customer satisfaction is based less on the quality of the deliverables than in traditional approaches. Certainly, the quality of the product still matters, but the quality of the working relationship can be just as important. In customer feedback surveys from Projects A1 and A2, an appreciation for the quality of the interaction with the development team was expressed alongside noted deficiencies in the final software deliverable. This suggests that when the customer feels that their needs are well understood and is pleased with the communication and interactions during the project, the likely result is a satisfied customer.

CONCLUSION

Quality assurance has been an integral part of agile development that has stemmed from process-inherent practices as well as practices for addressing specific quality goals. What makes quality assurance work so differently (McBreen, 2002) with agile projects is the way that quality goals are defined and negotiated throughout the project. From the seven project experiences that are discussed in this chapter, we were able to identify several common high-level quality goals and the different practices that were practiced to achieve them. Not every technique used was implemented perfectly in our projects, and not every technique was able to achieve a quality goal on its own. We attempt to address these areas of deficiency by dedicating a major portion of this chapter to the lessons that we have learned.

From the implicit suggestions in the lessons learned section for improving QA in agile projects, we feel the most important is: Actively attempt to capture and exploit informal communications. Our experiences have shown how valuable the information such as electronic correspondence, side discussions, and even code itself can be. When we used Storyboards in certain projects, we found that this way of capturing the informal communications between stakeholders helped new developers, as well as customers, get up to speed quicker and exposed difficult-to-predict issues. Moreover, in projects where informal communications were not captured, extra individual efforts were often made by team members to ascertain relevant information in order to understand requirements or complete programming tasks. Although the use of this particular living document is not a cure-all, the informal knowledge that it stores has helped us achieve a high level of software quality in functionality and usability.

For our own purposes, this chapter has also suggested that it is important to identify quality goals early on in the project, even though they may change. Not only does this keep the entire team mindful of these goals but it also allows for the planning of QA practices that will help us achieve them. In this way, QA can become agile—if a new quality goal is introduced in an iteration, appropriate QA practices can be selected and incorporated.

As the agile development group at Siemens Corporate Research, we are interested in identifying metrics for measuring software quality in agile projects. Although the success of QA practices in agile development projects is often measured by customer satisfaction, we recognize the need for measuring how agile processes and other factors influence software quality. We found that the specific ISO (ISO/IEC, 2003) metrics for measuring software quality were often vague, irrelevant, or unsuitable. With better metrics, we hope to make more concrete contributions to advance the understanding of quality assurance in agile projects.

REFERENCES

Beck, K. (1999). *eXtreme programming explained: Embrace change*. Addison Wesley.

Beck, K. et al. (2001). *Manifesto for agile software development*. Retrieved November 21, 2005, from http://agilemanifesto.org/

Brooks, F.P. Jr. (1987, April). No silver bullet: Essence and accidents of software engineering. *Computer Magazine, 20*(4), 10-19.

Gunaratne, J., Hwong, B., Nelson, C., & Rudorfer, A. (2004, May). Using evolutionary prototypes to formalize product requirements. Paper presented at *Workshop on Bridging the Gaps II: Bridging the Gaps Between Software Engineering and Human-Computer Interaction, ICSE 2004*, Edinburgh, Scotland.

Hwong, B., Laurance, D., Rudorfer, A., & Song, X. (2004, April). User-centered design and agile software development processes. Paper presented at *Workshop on Bridging Gaps Between HCI and Software Engineering and Design, and Boundary Objects to Bridge Them, 2004 Human Factors in Computing Systems Conference,* Vienna, Austria.

ISO/IEC 9126: *Software engineering—Product quality.* (2003). Switzerland: ISO.

Jacobson, I., Booch, G., & Rumbaugh, J. (1999). *The unified software development process.* Reading, MA: Addison Wesley Longman.

Martin, R. C. (2003). *Agile software development: Principles, patterns, and practices.* Prentice Hall.

McBreen, P. (2002). *Quality assurance on agile processes.* Software Craftsmanship Inc., Talks. Retrieved May 3, 2006, from http://www.mcbreen.ab.ca/talks/CAMUG.pdf

McConnell, S. (1996). *Rapid development.* Redmond, WA: Microsoft Press.

McConnell, S. (2004). *Code complete* (2nd ed.). Redmond, WA: Microsoft Press.

Nelson, C., & Kim, J. S. (2004, November). Integration of software engineering techniques through the use of architecture, process, and people management: An experience report. *Proceedings of Rapid Integration of Software Engineering Techniques, 1st International Workshop, RISE 2004, LNCS 3475* (pp. 1-10). Berlin & Heidelberg: Springer-Verlag.

Pichler, R. (2006, January 19). Agile product development: Going agile at Siemens communications. Presented at *OOP 2006*, Munich, Germany.

Schwaber, K., & Beedle, M. (2001). *Agile software development with Scrum.* Prentice Hall.

Smith, P. G., & Pichler, R. (2005, April). *Agile risks/agile rewards. Software Development*, 50-53. Retrieved May 3, 2006, from http://www.ddj.com/showArticle.jhtml;jsessionid=H1VRQ0BO1INWEQSNDBECKH0CJUMEKJVN?articleID=184415308

Song, X., Matos, G., Hwong, B., Rudorfer, A., & Nelson, C. (2004, November). People & project management issues in highly time-pressured rapid development projects. Paper presented at *EuroSun 2004*, Cologne, Germany.

Song, X., Matos, G., Hwong, B., Rudorfer, A., & Nelson, C. (2005, August). S-RaP: A concurrent prototyping process for refining workflow-oriented requirements. *Proceedings of the 13th IEEE International Conference on Requirements Engineering* (pp. 416-420). IEEE Conference Publishing Services.

Tai, G. (2005, May). A communication architecture from rapid prototyping. *Proceedings of the 2005 Workshop on Human and Social Factors of Software Engineering* (pp. 1-3). New York: ACM Press. DOI= http://doi.acm.org/10.1145/1083106.1083120

Turner, R., & Boehm, B. (2003, December). People factors in software management: Lessons from comparing agile and plan-driven methods. *The Journal of Defense Software Engineering.* Retrieved May 3, 2006, from http://www.stsc.hill.af.mil/crosstalk/2003/12/0312Turner.pdf

About the Authors

Lars Bendix is an associate professor at Lund Institute of Technology, Sweden, where he initiated the Scandinavian Network of Excellence in Software Configuration Management as a framework for collaboration between academia and industry. Software configuration management has been one of his main research interests for more than a decade. Bendix teaches the subject at university, has given several tutorials for industrial audiences, and has worked with many companies to improve their configuration management practices. He is also involved in his department's software engineering teaching that is based around the use of eXtreme Programming. He received his master's degree in computer science from Aarhus University, Denmark (1986) and his PhD from Aalborg University, Denmark (1996).

Eleni Berki is an assistant professor of software development at the Department of Computer Sciences, University of Tampere, Finland. Previously she was a researcher/principal investigator at the Information Technology Research Institute and assistant professor of group technologies in Jyväskylä University, Finland. Berki obtained her PhD in process metamodelling and information systems method engineering (2001) in UK. Her teaching and research interests include testing, security and trust, virtual communities, information and communication technologies, computational models, multidisciplinary approaches for software engineering, knowledge representation frameworks, and requirements engineering, whereon she supervises MSc and PhD students. She has worked as a systems analyst, a designer, and an IS quality consultant in industry, and has a number of international academic and industrial projects. She has been active in the development, delivery, and coordination of virtual and distance e-learning initiatives in collaboration projects in European and Asian countries. She has authored and co-authored more than 50 publications in world congresses, international forums, books, and journals and has given a number of talks in international conferences. Berki has been a professional member of the Institute of Electrical and Electronic Engineers (IEEE), the British Computer Society (BCS), and the United Kingdom Higher Education Academy within which she participates in organising and scientific conference committees, reviewing, and evaluation, and other collaboration projects. She has been a visiting lecturer in a number of UK Universities, the University of Sorbonne Paris-1 in France, and University of Crete in Greece.

Lindsey Brodie is currently studying for a PhD at Middlesex University, UK. She holds an MSc in information systems design. She edited Tom Gilb's latest book, *Competitive Engineering*. Previously she worked for many years for International Computers Limited (ICL), UK, carrying out technical project support, product support (operating system and database support), and business process and management consultancy. Brodie is a member of the British Computer Society and a Chartered Engineer (MBCS CITP CEng).

Yael Dubinsky is a visiting member of the human-computer interaction research group at the Department of Computer and Systems Science at La Sapienza, Rome, Italy, and an adjunct lecturer in the computer science department at the Technion-Institute of Technology. Her research examines the implementation of agile software development methods in software teamwork, focusing on software process measurement and product quality. She received her BSc and MSc in computer science and PhD in science and technology education from the Technion-Israel Institute of Technology. Dubinsky is a member of IEEE and IEEE Communications Society.

Barry Dwolatzky is professor of information engineering and director of the Information Engineering Research Programme at the University of the Witwatersrand, Johannesburg, South Africa. After obtaining his PhD from the University of the Witwatersrand (1979), he spent 10 years in Britain carrying out post-doctoral research at UMIST in Manchester, Imperial College in London, and at the GEC-Marconi Research Centre in Chelmsford, Essex. He returned to his alma mater as a senior lecturer in 1989. Dwolatzky's current research interests are in software engineering and in the use of geospatial information by utilities in developing countries. He is academic director of the Johannesburg Centre for Software Engineering (JCSE) at Wits University.

Torbjörn Ekman is a researcher at Lund Institute of Technology, Sweden, where he has been working within the LUCAS Center for Applied Software Research for the last five years. He is currently working in the PalCom integrated project in EU's 6th Framework Programme. His interest for software configuration management in agile methods started during the development of an eXtreme Programming course, which has now been running for four years. Other research interests include tool support ranging from refactoring aware versioning to compiler construction. Ekman received his master's degree in computer science and engineering (2000) and his PhD (2006) from Lund University, Sweden.

Eli Georgiadou is a principal lecturer of software engineering in the School of Computing Science, Middlesex University, London, UK. Her teaching includes comparative methodologies, evaluation of methods and tools, systems development, software quality management, and knowledge and project management. She has worked in industry in the UK and in Greece, and has extensive expertise in curriculum development and pedagogic issues gained in the UK, in Europe (primarily in Greece, Spain, and Finland), and further afield (such as in China, Egypt, and Vietnam). Georgiadou's research includes software quality, requirements engineering, information systems development methodologies, metamodelling, measurement, knowledge management, and process improvement but also in pedagogic issues, such as resource based open and distance learning. She has published over 90 refereed papers in journals and international conferences. She has organised and chaired a number of international conferences, workshops, and technology transfer initiatives. She currently coordinates the European Affairs and International Exchanges in her school and serves on various reviewing and scientific committees.

Tom Gilb has been an independent consultant, teacher, and author since 1960. He works mainly with multinational clients; helping improve their organizations, and their systems engineering methods. Gilb's latest book is *Competitive engineering: A handbook for systems engineering, requirements engineering, and software engineering using planguage* (2005). Other books are *Software inspection* (Gilb & Graham, 1993) and *Principles of software engineering management* (1988). His *Software metrics* book (1976, Out of Print) has been cited as the initial inspiration (IBM, Radice) for what is now CMMI Level 4. Gilb's key interests include business metrics, evolutionary delivery, and further development of his planning language, *Planguage*.

Beatrice Miao Hwong is a member of the technical staff in the Requirements Engineering Program of the software engineering department at Siemens Corporate Research in Princeton, NJ. She has led efforts most recently in process tailoring and requirements engineering in automotive, medical, and automation business units of Siemens. Hwong has a BS in electrical engineering from Tufts University, an MS from Cornell University, an MS CICE from University of Michigan, and a PhD in computer engineering from Rutgers University.

Maheshwar K. Inampudi is the lead IT architect for IBM's On Demand Workplace expertise location system (BluePages, and several other intranet applications). His additional responsibilities include the architecture and solution design for several of IBM's internal offerings as well as collaborating with the CIO's office and IBM Research in helping design applications using the latest SOA methodologies. He helps showcase IBM's emerging technologies such as WebSphere eXtended Deployment (XD) and IBM's IntraGrid Grid Computing Architectures. Inampudi holds a BS in computer science from Pune University and is currently pursuing a certification degree in advanced project management from Stanford University. Recent interests include leveraging emerging technologies, such as autonomic computing and grid computing.

Jim F. Kile is a PMI certified project management professional (PMP) and senior business area manager at International Business Machines Corporation working in Southbury, CT. He is responsible for managing a team of more than 135 individuals worldwide who develop, deploy, and maintain applications in support of IBM's internal corporate human resources. Kile holds a BBA in management information system from Western Connecticut State University (1989), an MS in information systems from Pace University (1995), and is currently pursuing his doctorate at Pace University. Throughout his career, he has created and piloted different project management and software development methodologies to improve the art and science of software development.

Monica McKenna is a member of the technical staff at Siemens Corporate Research, USA. She has more than 20 years experience in software design and development.

Scott Mark is an enterprise application architect for Medtronic, Inc, the world's leading medical technology company. His primary areas of expertise are the application of agile methods in the architecture and design of personalized information portals. He has applied agile methods in the context of dynamic content personalization for several large-scale, globalized systems. He has significant application development leadership experience, including disseminating development best practices and pattern-based design approaches in a large enterprise. He is skilled in the use of agile modeling and test-first development practices to enable lean project execution. Scott is a member of the Java Community Process (JCP) organization and was an early participant on the Expert Group for the JSR 170--Content Repository for Java Technology specification. He also has a strong background in technical writing and online information architecture and delivery.

Gilberto Matos is a member of the technical staff in the requirement engineering and rapid prototyping group at Siemens Corporate Research, USA. He has been involved in a number of internal and external software prototype and product development projects within Siemens over the last 8 years, mostly in technical lead roles. His research is centered on the methods and software support tools for faster and more accurate representation of user requirements in an executable form. Matos received his PhD in computer science from the University of Maryland at College Park.

Atif Memon is an assistant professor at the Department of Computer Science, University of Maryland. He received his BS, MS, and PhD in computer science in 1991, 1995, and 2001 respectively. He was awarded a Gold Medal in BS. He was awarded Fellowships from the Andrew Mellon Foundation for his PhD research. He received the NSF CAREER award in 2005. Memon's research interests include program testing, software engineering, artificial intelligence, plan generation, reverse engineering, and program structures. He is a member of the ACM and the IEEE Computer Society.

Ernest Mnkandla lectures in the School of Information Technology at Monash University, South Africa. He has submitted a PhD thesis at the School of Electrical & Information Engineering at the University of the Witwatersrand, Johannesburg, South Africa in the area of agile software development methodologies. He has lectured in this area and has presented several papers on agile methodologies and project management within Africa, Europe, and the Pacific Islands. Mnkandla completed a Btech (honours) in electrical engineering at the University of Zimbabwe (1992) and completed an MSc (Comp. Sc) at the National University Science and Technology in Zimbabwe (1997). His current research is in the adoption and quality assurance issues in agile methodologies. Mnkandla also does research in security policies for the implementation of wireless technologies.

Vagelis Monochristou holds a BSc in applied informatics from the University of Macedonia, Department of Applied Informatics (Thessaloniki, Greece), as well as an MSc in insurance and risk management from the CITY Business School, (City University, London, England). Since 2003, he is a PhD Candidate in the department of applied informatics in the University of Macedonia (Thessaloniki, Greece), and his research has been focused on the area of agile methods and the possibilities of their adoption in the Greek IT Market. Since 2000, Monochristou works as an IT consultant and has significant experience in business modelling, user requirements analysis as well as in software project management.

Christopher Nelson is an associate member of the technical staff in the software engineering department of Siemens Corporate Research, Princeton, NJ. His research has been in the areas of UI intensive software, agile processes, and global software development. Nelson received his BS in computer science and engineering from Bucknell University. He is currently attaining his master of software engineering degree from Carnegie Mellon University, Pittsburgh, PA.

Gergana Nikolova is a graduate student in computer science at the Technische Universität München. She recently worked within the rapid prototyping and requirements engineering group in the Software Engineering Department at Siemens Corporate Research. Niklova has contributed greatly towards the definition of a reference model for S-RaP (Siemens Rapid Prototyping)—an in-house prototyping process. Thus, her research focus has been in software development processes, particularly requirements engineering and rapid prototyping.

Orit Hazzan is an associate professor in the Department of Education in Technology and Science at Technion–Israel Institute of Technology. Her research interests are in human aspects of software engineering, particularly relating to agile software development and extreme programming development environments. She is coauthor (with Jim Tomayko) of *Human Aspects of Software Engineering* (Charles River Media, 2004). She received her PhD in mathematics education from the Technion–Israel Institute of Technology. Orit is a member of ACM and the ACM Special Interest Group on Computer Science Education.

Jörg Rech is a scientist and project manager of the Fraunhofer IESE. He earned a BS (Vordiplom) and an MS (Diplom) in computer science with a minor in electrical science from the University of Kaiserslautern, Germany. He was a research assistant at the software engineering research group (AGSE) by Prof. Dieter Rombach at the University of Kaiserslautern. His research mainly concerns knowledge discovery in software repositories, defect discovery, code mining, code retrieval, software analysis, software visualization, software quality assurance, and knowledge management. Rech published a number of papers, mainly on software engineering and knowledge management and is a member of the German Computer Society (Gesellschaft für Informatik, GI).

Arnold Rudorfer is the program manager for requirements engineering with worldwide responsibility at Siemens Corporate Research in Princeton, NJ. He has more than 12 years experience in product development and business consulting leading international projects. His main research interests are agile development techniques, requirements engineering as well as product management and marketing of software products. Rudorfer enjoys working with customers and his high-performance team to deliver solutions to Siemens businesses. Also, he is a certified Bootstrap and SPICE assessor.

Kerstin Siakas is an assistant professor at the Department of Informatics at the Alexander Technological Educational Institute of Thessaloniki, Greece since 1989. Her teaching includes software quality management, management information systems, and project management. She has developed and led large information systems projects in multinational companies in both Sweden and Greece. She has a PhD in software quality management from London Metropolitan University. Her research spans a range of issues in information systems quality, requirements, knowledge, and outsourcing management, in particular in cultural and organisational issues of the software development process, but also in pedagogic issues, such as technology based distance learning. Siakas has published around 50 refereed papers in different journals and international conferences.

Xiping Song is a senior member of the technical staff in the Software Engineering Department at Siemens Corporate Research, USA. In over a decade of working at Siemens, Song has been involved in many software development projects. He has worked as an architect for the Soarian project, designing the architecture for this key Siemens health service product. He designed the load-balancing strategy for deploying the product at the Siemens data center. Now, he is focusing on research on medical workflows. Song received a PhD from University of California at Irvine.

Grace Yuan Tai works as a member of the rapid prototyping and requirements engineering group within the Software Engineering Department at Siemens Corporate Research, USA. She is normally based in Princeton, but was working in Munich on a globally distributed agile project while working on this book chapter. She came to Siemens with a BS in computer science from the University of Michigan and has contributed to the research area of human and social factors in software engineering. Tai will continue to pursue her research interests as a master's student at RWTH Aachen starting October 2006.

Rajanikanth Tanikella is an associate member of the Technical Staff in the Software Engineering Department of Siemens Corporate Research, Princeton, NJ. Aside from software quality and testing related projects and prototypes for mobile devices, Tanikella has been involved in a number of agile development projects. He holds a BS in computer science from the State University of NY at Stony Brook and a MS in computer science from the New Jersey Institute of Technology. In his abundant free time, Tanikella does office work.

Qing Xie received her BS degree from South China University of Technology, Guangzhou, China (1996). She received the MS degree in computer science from the University of Maryland (2003). She is currently a PhD student in the Department of Computer Science at the University of Maryland. Her research interests include software testing, software maintenance, mutation techniques, and empirical studies. She is a student member of the ACM, the IEEE, and the IEEE Computer Society.

M. Vlachopoulou is an associate professor at the University of Macedonia, Department of Applied Informatics, Greece. Her studies include:

- Degree in business administration, Aristoteles University of Thessaloniki, Greece.
- Degree in law, Aristoteles University of Thessaloniki, Greece.
- Postgraduate degree studies in marketing, University of Mannheim, Germany, and MBS in business administration, Aristoteles University of Thessaloniki.
- PhD in marketing information systems, University of Macedonia, Department of Applied Informatics, Greece.

Vlachopoulou's main fields of research include: marketing information systems, e-business/e-marketing models, internet marketing plan, e-learning, new technologies and informatics in marketing, electronic commerce, ERP, and CRM systems, supply chain management (SCM) systems, knowledge management, e-supply chain management, e-logistics, virtual organization/enterprise modeling, and agile methods. She is the author and co-author of several books, mainly in the area of e-marketing, and has numerous publications in International Journals, Volumes, and International Conference Proceedings.

Bradley Wehrwein is an associate member of the Technical Staff in the Software Engineering Department at Siemens Corporate Research in Princeton, New Jersey, USA. He received his BS in computer science from the University of Massachusetts. His research interests include user interface, Web technologies, and agile software processes.

Index